Predictive Analytics in Healthcare, Volume 2

Transforming the future of medicine

Online at: https://doi.org/10.1088/978-0-7503-2317-8

About the Series

The series in Physics and Engineering in Medicine and Biology will allow the Institute of Physics and Engineering in Medicine (IPEM) to enhance its mission to 'advance physics and engineering applied to medicine and biology for the public good'.

It is focused on key areas including, but not limited to:
- clinical engineering
- diagnostic radiology
- informatics and computing
- magnetic resonance imaging
- nuclear medicine
- physiological measurement
- radiation protection
- radiotherapy
- rehabilitation engineering
- ultrasound and non-ionising radiation.

A number of IPEM–IOP titles are being published as part of the EUTEMPE Network Series for Medical Physics Experts.

A full list of titles published in this series can be found here: https://iopscience.iop.org/bookListInfo/physics-engineering-medicine-biology-series.

Predictive Analytics in Healthcare, Volume 2

Transforming the future of medicine

Vinithasree Subbhuraam
HerHeartCo Inc., Austin, TX, USA

IOP Publishing, Bristol, UK

ISBN   978-0-7503-2317-8 (ebook)
ISBN   978-0-7503-2314-7 (print)
ISBN   978-0-7503-2315-4 (myPrint)
ISBN   978-0-7503-2316-1 (mobi)

DOI   10.1088/978-0-7503-2317-8

Version: 20250501

IOP ebooks

British Library Cataloguing-in-Publication Data: A catalogue record for this book is available from the British Library.

Published by IOP Publishing, wholly owned by The Institute of Physics, London

IOP Publishing, No.2 The Distillery, Glassfields, Avon Street, Bristol, BS2 0GR, UK

US Office: IOP Publishing, Inc., 190 North Independence Mall West, Suite 601, Philadelphia, PA 19106, USA

This book is dedicated to the field of health technology and all the academicians, research scientists, patients, providers, healthcare professionals, and industry partners whose successful collaboration has birthed several innovative products and solutions that have transformed healthcare.

Contents

Preface

Predictive analytics (PA) is a branch of advanced analytics that uses techniques such as statistics, modeling, data mining, and artificial intelligence to analyze past and current real-time data to gain fast and accurate insights into several aspects of healthcare. Several books, publications, and courses online and offline provide comprehensive coverage of the technical implementation side of AI. However, few books present the application-oriented side of using PA and artificial intelligence (AI) in healthcare. The applications of AI in healthcare are evolving at an unprecedented pace, reshaping how we approach disease prevention, diagnosis, treatment, and medical education. Volume 1 of this book introduced PA and AI in layman's terms so that the information is accessible and digestible to providers, patients, and medical students who are now recognizing this paradigm shift in healthcare and becoming active participants. Volume 1 also presented applications and case studies in public health surveillance, women's health, telemedicine, and pervasive healthcare.

The time between the two volumes has seen exponential growth in AI in healthcare, with the widespread use of large language models (LLMs). This second volume dives deeper into cutting-edge advancements and more real-world applications. As more AI-powered ideas and innovations are surfacing rapidly, many from universities and other academic institutions, it is important for students/researchers/faculty to understand how to successfully convert their lab discoveries to real-life living and breathing products and solutions that help humanity. *Chapter 1* presents a typical path for academic entrepreneurship and addresses how entrepreneurs can trek through the 'valley of death' successfully.

As AI reshapes healthcare, it is revolutionizing how future medical professionals learn and adapt. *Chapter 2* explores the opportunities and challenges of using LLMs in rethinking how we deliver medical education.

Many patients have begun to explore holistic approaches to health and well-being and resort to complementary, alternative, and integrative medicine approaches for healing. However, these approaches suffer from widespread adoption and integration with Western medicine due to reasons such as a lack of sufficient quantitative evidence, lack of standardized protocols, and regulatory clearance issues. *Chapter 3* explores how AI can address these issues and showcases a few case studies.

The landscape of prostate cancer detection and care has rapidly evolved. It has moved from using conventional imaging and radical surgeries to an era of genomics, precision diagnostics, advanced imaging, and targeted treatments. *Chapter 4* presents the latest advancements in AI-driven applications across the continuum of the journey of a patient with prostate cancer.

Advancements in technology and the accumulation of large datasets are the perfect breeding grounds for ground-breaking innovations in precision medicine. *Chapter 5* describes the four most innovative AI-powered solutions that have disrupted this area—digital twins, multi-omics integration, AI-driven biomarker discovery, and personalized therapeutics.

Welcome to volume 2 of *Predictive Analytics in Healthcare*.

Acknowledgments

I sincerely thank Professor Dr Dhanjoo Ghista for providing me the opportunity to write two volumes of this book on predictive analytics in healthcare. My heartfelt and profound thanks are due to my Doctorate supervisors, Professor Dr Eddie Ng Yin Kwee, Nanyang Technological University, Singapore, and Professor Dr Rajendra Acharya, University of Southern Queensland, Australia. All three of them have been wonderful mentors and friends to me as I embarked on and continued my journey in biomedical data science. I still vividly remember conversations with them at Nanyang Technological University, which laid the foundation for my passion in this field. I am grateful for those precious times and the continued conversations over the last two decades!

I also wish to sincerely thank all the reviewers of this book proposal for their valuable suggestions.

I also want to express my gratitude to my colleagues at HerHeartCo. for cheering me on during this time and for the opportunity to build a life-saving interventional application for protecting women's heart health. It has kept my entrepreneurial fire burning and provided the fuel to write one of the chapters in this book on academic entrepreneurship.

Thanks are also due to Phoebe Hooper, Bethany Hext from IOP Publishing, UK, and Ed Mottram-Breeze from Institute of Physics and Engineering in Medicine (IPEM), UK, for their continued guidance while preparing this second volume. Thank you for your patience and understanding during the time between the two volumes. The time was fodder for various learning experiences in this field, which I have successfully (hopefully) translated into five in-depth chapters in this volume.

I want to express my heartful appreciation to my family, especially my sweet, now 11-year-old curious daughter, who has patiently supported and waited for me to finish this volume. Her 'Which chapter are you on, now?' and 'Are you done?' questions kept me on track. So thank you, my dear production manager!

Finally, thank you, Universe, for continuing to provide me with opportunities aligned with my passion and purpose—using technology to improve the health and wellness of the population. To the 20-year-old me who picked biomedical engineering and artificial intelligence as electives in her undergrad final year, this path has been a dream come true!

Vinithasree Subbhuraam
Austin, TX, USA
February 14, 2025

Author biography

Vinithasree Subbhuraam

Dr Vinithasree Subbhuraam is a digital health innovator, data-driven biomedical researcher and strategist, healthcare AI enthusiast, author, keynote speaker, and mentor with over 20 years of experience in the health tech industry. She is a 3× zero-to-one innovator who enjoys researching, developing, and implementing digital health products and solutions that address complex medical challenges and improve global health outcomes. Her professional experience has culminated in the development of several digital health products and solutions—an interventional application for women's heart health, software for pain management using postural therapy, an adjunct breast health monitoring wearable technology called the Cyrcadia Breast Monitor, systems for predicting fertility and onset of menopause using hormone markers, detecting the early onset of preeclampsia, and diagnosing polycystic ovary syndrome, and more. She has utilized predictive analytics for designing clinical decision support systems to detect diseases like carotid atherosclerosis, fatty liver, diabetes, epilepsy, and cancers in the thyroid, breast, ovaries, and prostate. She has an MS and PhD in biomedical engineering from Nanyang Technological University, Singapore, and an executive education in digital transformation in healthcare from Columbia Business School, USA.

She is currently the CTO of *HerHeartCo*. She is also the Associate Editor of the Q1 journal—*Computer Methods and Programs in Biomedicine* and an invited reviewer for several international journals. She has published over 100 journal papers, books, conference articles, and chapters. Her mission is to drive digital transformation in healthcare, leveraging her expertise in clinical research, data science, product development and management (idea generation, software and hardware development, clinical trials, IRB process, FDA submission), publications, fundraising, and stakeholder management.

Linkedin: https://www.linkedin.com/in/vinithasree-subbhuraam/

IOP Publishing

Predictive Analytics in Healthcare, Volume 2
Transforming the future of medicine
Vinithasree Subbhuraam

Chapter 1

From lab to life: navigating academic entrepreneurship to revolutionize health tech

Research is turning money into knowledge.
Innovation is turning knowledge into money.
—Geoff Nicholson, the former vice president of 3M

Universities and other academic institutions have vast intellectual resources and are naturally the breeding ground for future innovations. Acamedic researchers, faculty, and students and their work are critical to sustaining the wheel of innovation. As we navigate an AI-dominant world, more and more AI-powered ideas and innovations will surface rapidly. How can a student/researcher/faculty successfully convert their lab discoveries to real-life living and breathing products and solutions that help humanity? How can they navigate the 'valley of death' (VoD)—the gap between academic research and commercialization? These are some of the questions answered in this chapter. Tips from a successful academic entrepreneur and a case study of a highly successful academic startup are also included. The aim of this chapter is to inspire the next wave of academic entrepreneurs to bring more research from the lab to real life.

1.1 Introduction

One of the greatest joys of being a researcher is to experience a sense of excitement and fulfillment when the experiments to prove a hypothesis finally succeed or when an unexpected discovery surfaces. Another one, to most researchers, is the realization that their findings could potentially become a real-world product or solution. Let's say a research group has discovered a novel biomarker that can transform how cancer is diagnosed. Or a new implantable medical device based on advanced materials science research. How can a faculty, researcher, or student at an academic institution actively pursue the commercialization of these discoveries? How to translate scientific discoveries into tangible healthcare products or services? After all, not all academic discoveries are supposed to live in peer-reviewed scientific journal articles forever.

Universities and other academic institutions have vast intellectual resources, and they are naturally the breeding ground for future innovations. Academic entrepreneurship has recently grown in importance as universities and research institutions have increasingly and consistently recognized the need to transfer technology and knowledge from research to the real world. An academic entrepreneur is a professor or student from a university/school who sets up a business company to commercialize the results of his/her research. An academic startup is a new venture by university researchers (faculty and students), potentially outside university control. Academic spin-offs are companies created based on university-owned intellectual property (IP) and spun out of the universities.

Academic research plays a central role in the translational ecosystem. However, like traditional entrepreneurship, academic entrepreneurship comes with unique challenges. This is because there is generally no clear translational path to follow, and even if there is, the path varies from institution to institution and country to country. Section 1.2 details one of the most common paths.

Unfortunately, academic entrepreneurship also sits on one end of the VoD. The VoD is a crucial initial phase of new ventures—it is where substantial work has begun, but sufficient revenue has not yet been generated. The natural outcome is that these ventures run out of the initial investment. This is the phase where many academic startups or spin-offs fail. Sections 1.3 and 1.4 describe the VoD, detail the critical challenges researchers face when crossing the valley of death, and list some strategies to help them cross the same path.

Despite the challenges faced by academic entrepreneurs, several largely successful academicians have mastered the art of launching products out of academia into the research world. One such professor is Dr Robert Langer. Dr Langer is a David H. Koch Institute award-winning professor at the Massachusetts Institute of Technology. The Langer Lab conducts research at the interface of biotechnology and materials science. He has founded over 25 companies, holds over 1400 patents, and has licensed or sub-licensed his patents to over 400 biotech, pharmaceutical, chemical, and medical companies. He is also the co-founder of Moderna, the biotech startup that produced the vaccine for COVID-19. Other products based on his research include a growth hormone to treat dwarfism, a wafer to deliver chemotherapy drugs directly to tumors, and the Gliadel Wafer, a treatment for glioblastoma multiforme brain cancer. How did he do it? section 1.5 describes his approach.

Section 1.6 presents a case study of how an idea conceived by a student in an academic setting successfully transitioned from the research lab into a scalable startup. Final remarks are provided in section 1.7.

1.2 Typical roadmap of academic entrepreneurship

Academic entrepreneurship follows a structured path from the point of research discovery to a market-ready product. This section briefly describes the various steps in a typical roadmap (figure 1.1) in this process.

Initial Path of Academic Entrepreneurship

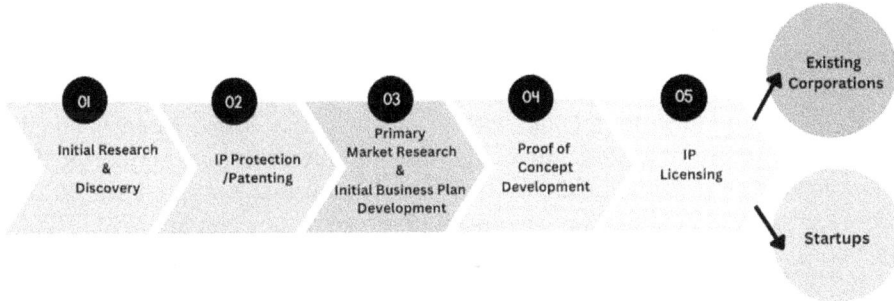

Figure 1.1. A typical initial path of academic entrepreneurship.

1.2.1 Initial research and discovery

Though academic settings are inherently creative places, research groups might suffer from the 'ivory tower syndrome.' It is a condition where students, faculty, and researchers are more focused on research and theoretical development and less connected to how their work can exist within the practical realities of the market and the needs of potential customers. They might lack understanding of how to effectively commercialize their research and innovations, staying 'in the ivory tower' without engaging with the real-world challenges of entrepreneurship. There is more pressure to publish research papers and secure grants for further research. They also lack business acumen in marketing, finance, and product development and can resist industry partnerships. So, the first critical consideration for a researcher, student, or faculty member who wants to embark on the journey of becoming an academic entrepreneur is to work on ideas that have scientific feasibility and potential applications in the real world. It is also the perfect time to cultivate an entrepreneurial mindset by blending scientific curiosity with business strategy. The next step is conducting preliminary research that leads to an invention or discovery worth commercialization.

1.2.2 Intellectual property (IP) protection

A research discovery is of value only if it is protected so no one else can use it without permission. So, the next step is identifying possible intellectual property, which can be patents, trademarks, copyright, or trade secrets. Care must be taken around public disclosure when a patent process is in progress. A public disclosure is information that provides sufficient details to allow someone to reproduce the invention. For academics, public disclosure could happen in various ways: printed publications such as book chapters, conference abstracts, posters, proceedings, journal articles, theses, emails, or other correspondence to people without stating that the information is confidential, grant proposals to federal agencies, etc. If a meeting is needed with someone outside the university to discuss the invention, it is important for both to sign a non-disclosure agreement (NDA). Even though academics enjoy sharing the excitement around their research findings, protecting the invention is key if commercialization is on the horizon.

In the university setting, the IP protection process starts with an internal disclosure to the technology transfer office (TTO). The inventor (faculty/student) and the TTO will then conduct a prior art search to determine if anything else is close to the invention in the public literature or other published patents. The TTO will take time to evaluate some metrics of the invention (such as market potential, commercialization potential, etc. Another key element is the internal budget of the TTO, which generally limits the amount of disclosures they can approve for patent conversions. Suppose the TTO does not approve and patent the invention. In that case, the inventors can choose to analyze the viability of the business opportunity, hire their patent attorney, and get the patent assigned to themselves. The advantage is that there is no need to license the technology from the university, and they are free to build the business. If the TTO decides to protect the invention, the university will own the patent, and a startup or spin-off should license the technology from the university. At this stage, it is also important to decide on the inventorship proportions so that when the technology gets licensed, IP income distribution becomes seamless. There are more nuances to the patent filing process, which are beyond the scope of this chapter.

1.2.3 Market analysis and initial business plan development

Conducting a thorough market analysis to understand if there is a demand for the invention in the market is critical to de-risking the business opportunity. If the primary market research results do not indicate a resounding yes for the invention, it is time to analyze how the invention can be made better to meet the market need. Adding more features, strategizing pricing better, or combining the invention with another to add more value could be a few ways to improve the product-market fit. If none of the potential customers want the concept, the product can be reworked and tested again in the market. After all efforts, if the market research outcome is a no, it might be time to give up, publish papers around the invention, and try again with a new invention or discovery. Early market analysis is key and can save the inventor time and resources.

The TTO can be a good source of information during the market analysis phase. They may have subscribed to marketing reports that indicate the market size and growth, and competitors. Another good place to learn more is the National Science Foundation's Innovation Corps (I-Corps) nodes and centers at universities. They have built a community of researchers and business advisors to provide education for assessing the commercial viability of research findings coming from academic labs.

After a successful primary market analysis, the next step is to establish the value proposition—value is the intangible and tangible benefits of the invention to customers. A value proposition statement reflects how the invention solves the identified problem for the target customers, highlights the key benefits, and identifies the differentiating factors that set the solution apart from what the competitors offer. Quantifying the value and market will help determine the pricing strategy—whether it is a one-time fee, recurring annual license, monthly/annual subscription model, etc. It is also time to consider how the product/solution will be introduced to the market—via influencers, social media, paid advertising, or several strategies. At every step, it is important to document the decisions with supporting evidence (such as marketing reports, customer interview analyses, etc), just like how a scientific paper has references. This simple but powerful step will increase the business plan's credibility with the investors.

The final step is financial analysis. Studying how much money it takes to turn the invention into a marketable product is important. This analysis is highly dependent on the product. A health tech software application will have a very different timeline and cost from a drug development process. The most important point here is that the investment it takes for the invention to become a product must be justified by its potential for making revenue in the market. The Founders Network has some great material on this topic (The Founders Network 2025). Upmetrics offers a financial statements template (Upmetrics 2025).

1.2.4 Prototype or proof of concept (PoC) development

For university startups, developing a PoC or prototype before obtaining an IP license could be beneficial for the following four key reasons:

(1) It allows for better validation of the invention's feasibility and market potential and shows that the technology has real-world applications beyond just being a theoretical idea. This, in turn, strengthens the case for licensing the IP.
(2) It de-risks the invention, makes the licensing process more attractive to potential investors, and increases the chances of a successful startup.
(3) It helps make informed licensing negotiations with the University's TTO.
(4) It helps to test the invention with potential users and identify potential market challenges and opportunities early on.

The PoC stage involves de-risking the invention and getting out of the early stages. Here, steps are taken to address risks associated with technology, regulatory, market, and management teams. The result of a PoC is not a finished product but the technological evidence that convinces investors that the invention will perform as intended in the real world. For example, it can be as simple as launching a minimum viable product (MVP) software platform for a short period to collect marketing data and user feedback, remove the platform from public use, and use the real data to make the product better. The results of PoC will provide an opportunity for the academic research facility to pivot, as necessary, early on. This step requires funding; sometimes, the universities or regional business development communities can help. Other options are federal programs such as the Small Business Innovation Research (SBIR)/Small Business Technology Transfer (STTR) programs (SBIR 2025) or Proof of Concept Centers (POCC 2025). Raising funds through angel investors is another option.

1.2.5 IP licensing

At this point, a market-ready invention has been determined, sufficient evidence has been collected via PoC, and provisional or non-provisional patent applications have been filed. Now, let's assume a startup or an established corporation is interested in licensing the IP.

License agreements with existing corporations: Corporations are generally interested in licensing IP if it is a one-time product or an improvement to an existing product or process. The company aims to improve its current process or product, expand its product offerings, or table the IP and strategically block others from using it. If a corporation is interested, the TTO will present a set term sheet and agreement that generally details how the TTO can recover the patent fees from the corporation and fees/royalties that are to be paid by the corporation based on sales of the invention. Sometimes, the corporation might want the invention not to use it, but to just block the competition from using it. Or their drive for commercialization might wane, and the invention could remain on the shelf. In such a scenario, no future royalty will be returned to the university. To address such scenarios, the TTO might add a contractual clause that states that the IP goes back to the university if the corporation doesn't make significant progress on the commercialization front within a time period.

License agreements with startups: If the invention is a platform technology with several applications, is disruptive and revolutionary, and addresses a new or a long unmet need, it is better suited to be commercialized via a new startup formation. The inventors (faculty members and students) could be this startup's employees or shareholders while still being the university's employees or students. However, such a situation can result in a conflict of interest for the students and faculty members as they balance their interests in the startup and the university. To address this, many universities now have predetermined deal structures for startups that mention how much equity, royalties, or other terms are assigned to the university in exchange for an exclusive license of the IP.

Suppose the IP is licensed to a startup. In that case, the next steps will be quite similar to those in a traditional entrepreneurship route—team formation, fundraising, product development, product validation (e.g., clinical and market validation via pilot studies, clinical trials, real-world feedback, etc), regulatory clearance (if needed), partnership development, product launch, marketing, sales, scaling, and sustaining growth. Many books (Shane 2004, Blank 2005, Ries 2011, Blank and Dorf 2013, Blumberg 2013, Horowitz 2014, Thiel and Masters 2014, Marcolongo 2017, Gooneratne *et al* 2021, Micalo 2022, Meyers 2023) and articles (Burkholder and Hulsink 2022, Sieg *et al* 2023, Engzell *et al* 2024, Glover *et al* 2024, Lim *et al* 2024) have been written to demystify these areas for new and seasoned entrepreneurs.

The next two sections discuss the VoD, a phase almost everyone venturing into the startup world will face. It indicates the series of obstacles and challenges that stifle the progress of transforming the invention into a real-life marketable and sellable product.

1.3 The valley of death

The term 'valley of death' was popularized by the National Research Council (NRC) report a decade ago to describe the phase where research on weather satellites struggled to advance to the point of delivering practical benefits for society (National Research Council 2000). There have been several definitions for VoD. However, the tech startup VoD has generally been widely recognized as the phase where fundamental discoveries and translation efforts are complete, substantial resources have been invested, and operations have commenced. However, revenue generation is still insufficient, and the technology is too premature to attract significant investment from venture capitalists. It is a myth that tech startups will encounter only a single VoD during their journey. Startups and even mature tech ventures will face multiple VoDs throughout their life cycles—as their science progresses from the lab to PoC, proof of business, proof of market, and proof of scale.

Specifically in health tech, VoD refers to the metaphorical valley that swallows research before it becomes a real-world clinical application. Biomedical research aims to deliver new treatments, diagnostics, and preventive strategies. However, despite significant advances in biomedical innovation and research, these advances have not significantly transformed into clinical solutions. Even when innovations do reach the market, patient access—especially in low-resource settings—remains a challenge. Overcoming this VoD requires a well-connected ecosystem involving data science and medical students, academicians, physicians, scientists, industry partners, funding agencies, investors, and regulatory bodies. Besides, several challenges need to be strategically addressed. Figure 1.2 illustrates a VoD cause assessment infographic that lists the possible challenges entrepreneurs face in the eight major areas of building a company. The issues listed in the infographic are self-explanatory, and a few or many of those might be well recognizable to startups stuck in this valley. A known issue is easier to address than an unknown issue. The infographic can be used as a checklist to understand why a startup is stuck in a valley and take the necessary steps to move forward.

Academic spin-offs or startups face some additional unique challenges, such as the following:

- *Time commitment:* Balancing academic responsibilities with entrepreneurial pursuits is one of the key challenges academic entrepreneurs face.
- *Funding:* Academic research, generally considered early-stage innovations, may not attract investors due to high risks and uncertain returns. Many grants focus on early-stage discovery but rarely support commercialization.
- *Regulatory hurdles:* Many academic teams lack knowledge of regulatory requirements and understanding of compliance frameworks such as FDA, HIPAA, GDPR, etc. If they innovate without considering these frameworks, there could be delays in getting clearance due to a lack of requirements such as adequate documentation, etc.
- *Technical challenges:* Scaling from small-scale experiments to large-scale production requires significant technical planning, and academicians might not have the relevant experience to understand or infrastructure to support scaling.

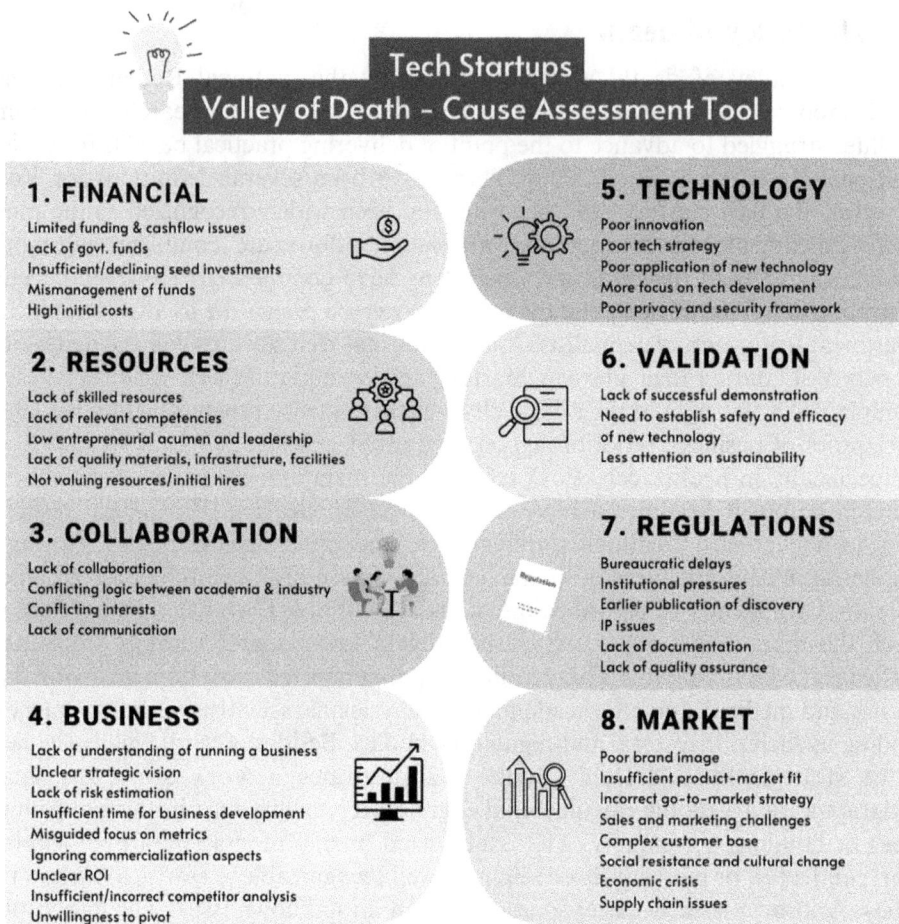

Tech Startups
Valley of Death – Cause Assessment Tool

1. FINANCIAL
Limited funding & cashflow issues
Lack of govt. funds
Insufficient/declining seed investments
Mismanagement of funds
High initial costs

2. RESOURCES
Lack of skilled resources
Lack of relevant competencies
Low entrepreneurial acumen and leadership
Lack of quality materials, infrastructure, facilities
Not valuing resources/initial hires

3. COLLABORATION
Lack of collaboration
Conflicting logic between academia & industry
Conflicting interests
Lack of communication

4. BUSINESS
Lack of understanding of running a business
Unclear strategic vision
Lack of risk estimation
Insufficient time for business development
Misguided focus on metrics
Ignoring commercialization aspects
Unclear ROI
Insufficient/incorrect competitor analysis
Unwillingness to pivot

5. TECHNOLOGY
Poor innovation
Poor tech strategy
Poor application of new technology
More focus on tech development
Poor privacy and security framework

6. VALIDATION
Lack of successful demonstration
Need to establish safety and efficacy
of new technology
Less attention on sustainability

7. REGULATIONS
Bureaucratic delays
Institutional pressures
Earlier publication of discovery
IP issues
Lack of documentation
Lack of quality assurance

8. MARKET
Poor brand image
Insufficient product-market fit
Incorrect go-to-market strategy
Sales and marketing challenges
Complex customer base
Social resistance and cultural change
Economic crisis
Supply chain issues

Figure 1.2. A VoD cause assessment infographic that lists the possible challenges entrepreneurs face in the eight major areas of building a company.

- *Limited business acumen:* Most academic researchers lack business skills and struggle with market research, business strategy, and fundraising.
- *Interdisciplinary collaboration:* Enabling successful collaboration between researchers, engineers, clinicians, and business experts is often challenging.

Addressing these challenges is imperative as the pace of startup growth has increased. The next section lists strategies for building bridges to cross the VoD in various aspects of startup management.

1.4 Strategies to cross the valley of death

Gbadegeshin *et al* (2022) studied over 128 scholarly materials and empirical data from 30 startups (in AI, virtual and augmented realities, the internet of things, medical, and cleantech industrial sectors). They observed that the most commonly cited VoD model was the one by Markham (2002). It depicts VoD as the gap between the availability of resources for R&D and the resources for commercialization. However, none of the models proposed how to prevent or flatten the VoD. So, they proposed a new and comprehensive model to cross the VoD for high-technology-based startups in this paper.

Though academic spin-offs/startups are ideally supported by parent organizations (universities), can participate in more government-supported financial programs and grants, and have a long history of scientific evidence supported by various prior research projects in the universities, they still face several issues, just as how traditional startups do. The following subsections present strategies that various startups use to cope with the VoD. Some of these strategies were shared by the companies interviewed by Gbadegeshin *et al* (2022), and some were gleaned from other related publications.

1.4.1 Where to look for the money?

Academic entrepreneurs must combine multiple funding sources to navigate any funding-related issues. Some non-dilutive or minimal dilutive funding options are listed below.

1. Non-dilutive government grants such as SBIR and STTR that support early-stage R&D and commercialization in phases. Other options include NSF's I-Corps program grants and translational research grants such as ARPA-H (ARPA 2025) and BARDA (BARDA 2025).
2. University Innovation funds and PoC grants such as Stanford SPARK (SPARK 2025).
3. University-affiliated startup incubators and accelerators. Examples include MIT Sandbox, Stanford StartX, and Harvard i-Lab.
4. Philanthropic and non-profit grants such as the ones offered by the American Heart Association and Gates Foundation.
5. Angel investors, angel networks, and venture capital who support pre-seed and seed ventures.
6. Strategic corporate R&D partnerships (healthcare, pharma, medical device companies) where the corporation provides investment.
7. Crowdfunding, where the public invests in startups in exchange for equity.
8. Revenue-based financing, where investors provide capital without taking equity in exchange for a percentage of future revenues.
9. Small business low-interest government-backed loans.

Figure 1.3 shows some commonly used strategies by companies to ensure that they operate within their means. The operating mantra, especially in the early stages, should be to spend cautiously. Some startups also include some features of the offering that could bring in revenue early on. For example, a digital health wellness platform can have a supplement line as one of its revenue streams while the core platform is built and tested.

Finances and funding

Look for both local and global public funding programs

Use early sales as a financing source

Analyze components that can raise leverage in funding negotiation (patents, R&D partnerships. etc.)

Prepare well for stakeholder meetings

Focus on decent investment offers instead of waiting for perfect deals

Avoid exorbitant salaries for executives and optimize spending on resources needed to run daily operations

Focus on securing sufficient funding to cover the essential operations

The main financial strategy of most of the interviewed startups - "lower costs, increased income"

Figure 1.3. Crossing the VoD—some common strategies to address financial issues.

1.4.2 Building the team

A strong team is one of the most important factors contributing to a startup's success. Teams can make or break growth. Forming a startup with the right co-founders, with at least one co-founder with entrepreneurial experience and one with technical expertise (especially for tech startups), will propel growth. University innovation hubs, incubators, and professional groups such as LinkedIn are excellent places to look for co-founders with complementary skills. Disagreements on long-term vision and culture misalignment (one wants to go slow and steady, and another wants to take an aggressive route) can cause conflict and low morale. So, it is also important to have an aligned vision and clear discussions early to define roles, equity, and responsibilities. Many bootstrapped startups have found solace in reducing the spending budget by hiring contractors, freelancers, and interns first before committing to full-time hires. However, hiring the right resources at any given time who align with the company's vision and mission is important. Once the right resources are in place, it is important to employ strategies to reduce the high employee turnover that is usually common in early-stage startups. Care must be taken to retain talent by keeping their morale high, setting realistic expectations of them, and providing incentives other than monetary. In a world where remote working has become more common, addressing any communication and collabo-ration issues distributed teams face is critical. Figure 1.4 presents a few other strategies for hiring and managing the right resources.

1.4.3 Collaborating across disciplines

Driving translational research from lab to real life requires a new collaboration model among various stakeholders. A health tech startup's stakeholders generally will include engineering and medical students/researchers, academicians, scientists, physicians, business experts, marketing and sales experts, engineering teams (hard-ware and software), industry partners, and legal and regulatory personnel. The first step in finding and collaborating with such varied experts is to build or utilize solid networking ecosystems. Figure 1.5 lists a few strategies for networking and collaboration.

One of the biggest bottlenecks to digital health innovation is not collaborating and co-creating with healthcare practitioners (HCPs). More often than not, it is still common to call a digital health product 'backed by clinicians' when, in reality, HCPs are often only 'passively consulted' by digital health innovators. Even though HCPs better understand the healthcare system, patient needs, medical protocols, and clinical workflow, they are mostly excluded from the active co-creation process. No wonder even fully mature solutions struggle to get adopted by HCPs. However, it is not easy to achieve active co-creation with busy HCPs in the current environment where many physicians are experiencing burnout.

Here are some ways to help achieve better co-creation with HCPs:

1. **Incentivization:** Financial incentives alone are not sufficient. The pressing issue for HCPs is time, and offering solutions to optimize their workflow efficiency and time and solutions that improve clinical outcomes could help. And sometimes, asking an HCP what they need in exchange for their valuable time and expertise is a great start.

2. **Awareness and education:** Most HCPs are unaware of the plethora of digital health tools available. The clinical curricula must be updated to include details on these tools not only to enable them to prescribe them to their patients but also to allow them to share their experience with the innovators.

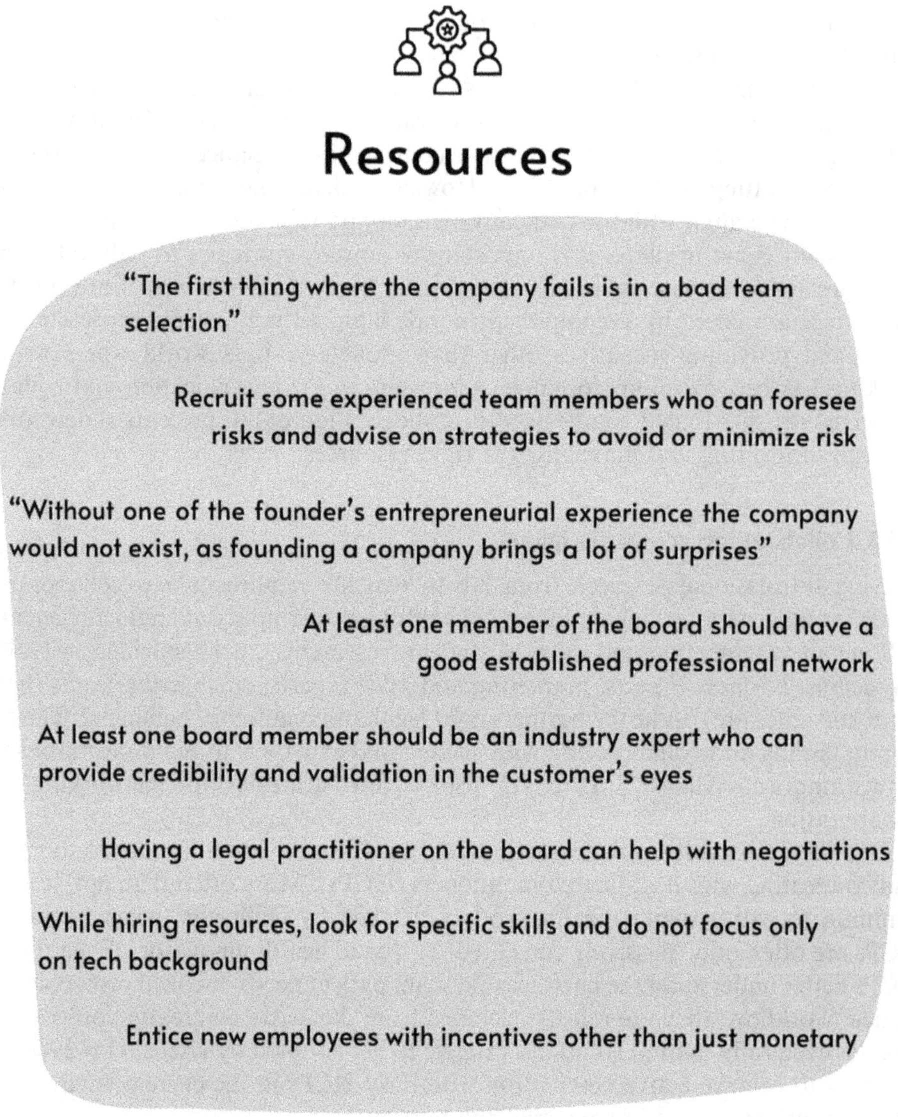

Resources

"The first thing where the company fails is in a bad team selection"

Recruit some experienced team members who can foresee risks and advise on strategies to avoid or minimize risk

"Without one of the founder's entrepreneurial experience the company would not exist, as founding a company brings a lot of surprises"

At least one member of the board should have a good established professional network

At least one board member should be an industry expert who can provide credibility and validation in the customer's eyes

Having a legal practitioner on the board can help with negotiations

While hiring resources, look for specific skills and do not focus only on tech background

Entice new employees with incentives other than just monetary

Figure 1.4. Crossing the VoD—some common strategies for forming a great team.

3. **Evidence:** Even if digital tools are available and education is provided, lack of evidence prevents HCPs from trying them. Most companies, especially those developing solutions with an AI component, must take significant steps to establish clinical evidence and robustness (e.g., regulatory filings, clinical trials, pilot studies, real-world evidence studies, etc) to get HCP buy-in.
4. **Integration:** Providers are generally not too excited to utilize several fragmented solutions for the same specialty in medicine. Solution A for one aspect and B for another will not be efficient for anyone. Startups should consider whether digital health solutions can be connected, contextualized,

Networking/Collaboration

Recruit team members, employees, advisors, and board members via personal networks

Use personal networks to get suppliers, first customers, collaborators, and investors

Use existing connections with listed industry giants to get early sales

Broaden the existing network for further progress and sustainability

Attend expos, conferences, innovation summits, exhibitions to build solid network ecosystems

Spinoff startups can make use of their university networks - and can therefore maintain good relationships with public organizations

Figure 1.5. Crossing the VoD—some common strategies to build an effective network.

and communicated through efficient collaborations and partnerships, accelerating the adoption among HCPs and users. In such cases, the age-old proverb should read, 'If you want to go slow, go alone. If you want to go far, go together.' Gathering HCPs' input on what they want for their practice regarding integrated solutions would be beneficial.

5. **Mandation:** Regulators can mandate HCP participation in the innovation process as a critical component of their approval checklist to establish the solution's clinical safety and efficacy.

1.4.4 Learning to run a business

Suppose the invention is to be commercialized by an academic startup. In that case, it is important for the inventors and/or their business counterparts to properly set up business registration, shareholder documents, and other legal and operational paperwork. A legal structure and business name should be first chosen to register a company. The next steps are to apply for tax IDs, file Articles of Incorporation/Organization, and create operating agreements or Bylaws. An important consideration here is to figure out where to register the company. If the startup has co-founders, a Founder's Agreement should be in place that outlines the roles and responsibilities of each founder, equity distribution, vesting schedules, and exit terms. A shareholder agreement outlining how shares are transferred, bought, or sold should also be created. There should also be a cap table that outlines the ownership stakes in the company, detailing the number of shares or ownership percentages held by each shareholder, founder, investor, and employee. If the company has a board of directors, board resolutions should be written to formalize decisions, such as issuing new shares, approving significant contracts, or making strategic decisions. As mentioned earlier, academic startups will have a licensing agreement that details the rights to commercialize the IP, any royalties to be paid, and any restrictions on use. Care should be taken to ensure that all intellectual property (e.g., patents, copyrights, trademarks, trade secrets) created by the inventors is assigned to the company.

Before sharing the ideas with potential collaborators or even during the hiring phase, the key document that needs to be in place is the NDA. It helps protect the IP and business secrets. If the business handles consumer data (e.g., a SaMD (software as a medical device platform) for wellness), care should be taken to follow any privacy laws and the platform should include privacy policies and terms of services.

Most academic entrepreneurs find this area overwhelming, but these documents are necessary to ensure the business is appropriately structured and protected. This is an area worth spending those initial dollars on. So, hiring legal and financial advisors to create the documents is good practice. Figure 1.6 highlights some strategies suggested by startups interviewed by Gbadegeshin *et al* (2022).

Besides creating solid business documents, a crucial but often overlooked aspect of a startup is raising the company's valuation. Startup valuation is the process of determining a startup's monetary worth at a given stage of its growth. Increasing valuation is important for fundraising, investor negotiations, acquisitions, financial planning, corporate and clinical partnerships, and successfully scaling the business. Unlike mature companies, startups often lack revenue or profits, so valuation is based on the following areas:

- **Market potential:** Startups operating in large, fast-growing markets generally have higher valuations. Understanding the total addressable market (TAM) and scalability potential is important. For example, a startup targeting a $100B+ healthcare AI market will likely get a higher valuation than one targeting a $500M niche segment.

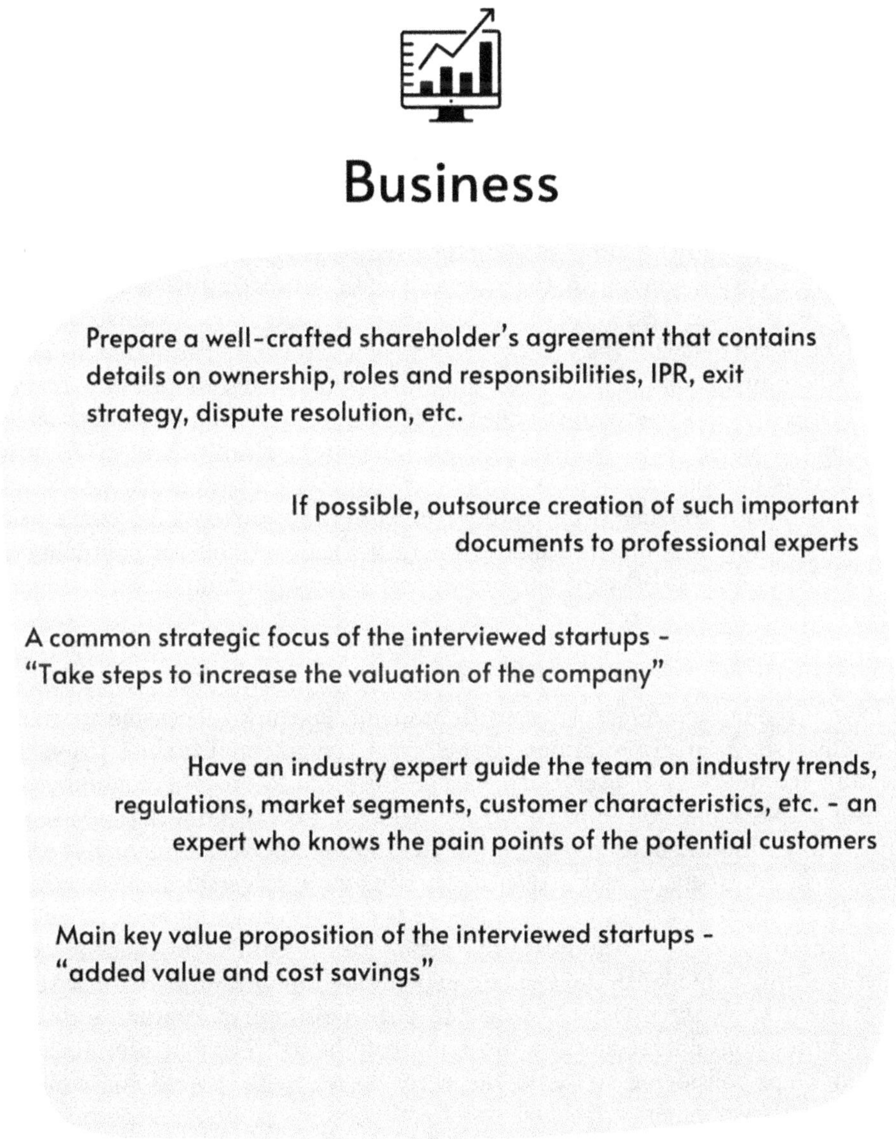

Business

Prepare a well-crafted shareholder's agreement that contains details on ownership, roles and responsibilities, IPR, exit strategy, dispute resolution, etc.

If possible, outsource creation of such important documents to professional experts

A common strategic focus of the interviewed startups – "Take steps to increase the valuation of the company"

Have an industry expert guide the team on industry trends, regulations, market segments, customer characteristics, etc. - an expert who knows the pain points of the potential customers

Main key value proposition of the interviewed startups – "added value and cost savings"

Figure 1.6. Crossing the VoD—some common strategies to address business-related issues.

- **Market traction:** Investors value startups that can demonstrate the presence of demand signals for their products. Demand signals can be captured by showing revenue growth, user adoption/interest/engagement metrics, paid pilot projects, long-term contracts with interested business entities, letters of intent, and customer pre-orders. For example, a digital health platform with ten signed provider contracts will be valued higher than one with no real-world adoption metrics.

- **Strength of IP:** Strong patents and the use of proprietary technology naturally increase a startup's valuation, credibility, and investor confidence.
- **High-caliber teams and advisory board:** A strong co-founding team with proven expertise and an advisory board with top-tier advisors, key opinion leaders, healthcare providers, and other industry experts increase valuation.
- **Strategic partnerships:** Institutional (universities and research organizations) and industry partnerships demonstrate the product-market fit and increase credibility and valuation. Having distribution partnerships in place for when the product launches in the market is also an excellent indicator of the product's long-term success.
- **Competitive edge:** If the startup operates in a highly competitive environment, valuation decreases if it does not differentiate itself sufficiently from its competitors. A competitive edge can be achieved in many ways, such as with unique selling features, deep tech use, regulatory clearance, or even simpler aspects like superior UI/UX.
- **Non-dilutive funding:** Securing government grants and winning competitions or securing non-dilutive funding by other means establishes credibility and visibility and de-risks the venture without equity dilution.

1.4.5 IP rights and technology validation

IP: One of the most common issues with academic startups is licensing technology from universities to corporations to fast-track commercialization. Universities generally do not license everything easily. Firstly, the lack of communication between corporations and university TTOs leads to no one knowing the existence of inventions, and the burden falls on the TTO to market their IP, which is a time-consuming and expensive task. Also, negotiating with universities and waiting for layers of approval processes to be completed is a time-consuming process unfavorable to corporations. To address this issue, the inventors can market the IP themselves via conferences and other networking opportunities to ensure that corporations know the existence of their invention/discovery. Another option is to hire business consultants and let them market the IP. Finally, most academic inventions are considered to be in the early stages, with a need for additional research and development steps. One way to mitigate this is for the inventors to start their own company by licensing the technology from the university, raise some seed money to de-risk the technology by doing PoC and feasibility studies, and then approach a corporation for a partnership to continue the commercialization process. If a corporation is interested in licensing, it is important to thoroughly understand the tech transfer policies and scrutinize the existing patents and preliminary market research data before making a deal (figure 1.7).

Technology validation: It is important to ensure the invention is technically, clinically, and commercially viable to make it attractive to investors and industry partners. The first step is to identify what needs to be validated. In the technical sense, testing whether the technology can perform as expected in real-world conditions is important. In the clinical/regulatory sense, the inventors should first

Technology & IPR

If it is a university startup or spinoff, closely look into the potential issues relating to university tech transfer policies and terms

Consider the relevance of timing for tech transfer or commercialization process due to possible challenges associated with IPR

Before making a licensing deal with universities, ensure that the patent, software, and market research data are worthy

"Despite a large number of patents, highly skilled people were seen as the most important in a company"

Figure 1.7. Crossing the VoD—some common strategies to carefully license university IPs.

understand if the invention needs regulatory clearance and if it should meet industry standards such as ISO. On the business end, the key question is whether the customers will pay for it. The next step is to develop a minimum viable product and conduct preclinical studies to validate the technology before getting the funding for large-scale trials or real-world studies. Successful preclinical or small cohort studies can lead to PoC studies in real-world settings that can be conducted in partnerships with hospitals and corporations. For regulatory clearance, it is important to engage with regulator agencies early and understand the pathway required for the invention to enter the market legally. These agencies can provide the best guidance in sample size requirements for acceptable studies and advise on the framework for AI-powered health tech innovations.

1.4.6 Market and customer acquisition

Conducting user research and engaging end-users early in the development process is key. Steve Blank, a serial entrepreneur from Stanford University, proposed a market analysis approach that has been adopted by the National Science Foundation's I-Corps program. He proposes that the inventors conduct this primary market research approach as they understand their invention better than any third party. His approach is to use the customer discovery data to develop a market hypothesis, test that hypothesis with at least 100 interviews with potential customers, and validate the customer base. Then, refine the hypothesis based on the results as one builds the product. Some other valuable marketing strategies are shown in figure 1.8.

In health tech startups, customer acquisition is challenging due to regulatory hurdles, long sales cycles, and the involvement of multiple decision-makers (patients, providers, payers). A successful strategy mainly involves a mix of B2B (providers, pharma, payers) and B2C (patients, caregivers, fitness providers, etc) approaches. For any approach to work, it is first important to define an ideal customer profile based on pain points, needs, and budgets, then look for early adopters open to innovation.

If the target customers are consumers, digital marketing strategies (SEO and content marketing, influencer marketing, webinars, podcasts, white papers, case studies, and research-backed blogs) could help. Offering free trials and affordable subscriptions could also be attractive to consumers. Gamification and performance-based pricing (pay-per-success for health improvements) are great strategies to reduce consumer churn and keep engagement metrics high.

For B2B models, where the target customers are mostly healthcare providers, the easiest way for buy-in is to co-create with them and launch through their networks. In any global market, physicians, hospitals, and insurers embrace adoption if the product shows strong technological validity and regulatory compliance. Once these trust elements are established, using ROI-driven messaging such as 'product reduces hospital readmissions by 30%', etc, publishing in healthcare journals, getting media coverage, listing in digital health marketplaces, and speaking to Chief Medical Officers and Innovation Directors of hospitals are other strategies that can help establish partnerships and win contracts. Many providers and hospitals trust word-of-mouth more than ads. Implementing referral and affiliate programs that offer discounts, credits, and incentives for referrals and engaging health influencers and physician networks are key.

1.4.7 Mastering sales

Entrepreneurs should adopt early sales strategies to gain paying customers, generate revenue (that can ideally be reinvested), and attract investors. Besides following customer acquisition strategies mentioned in section 1.4.6, startups can follow some

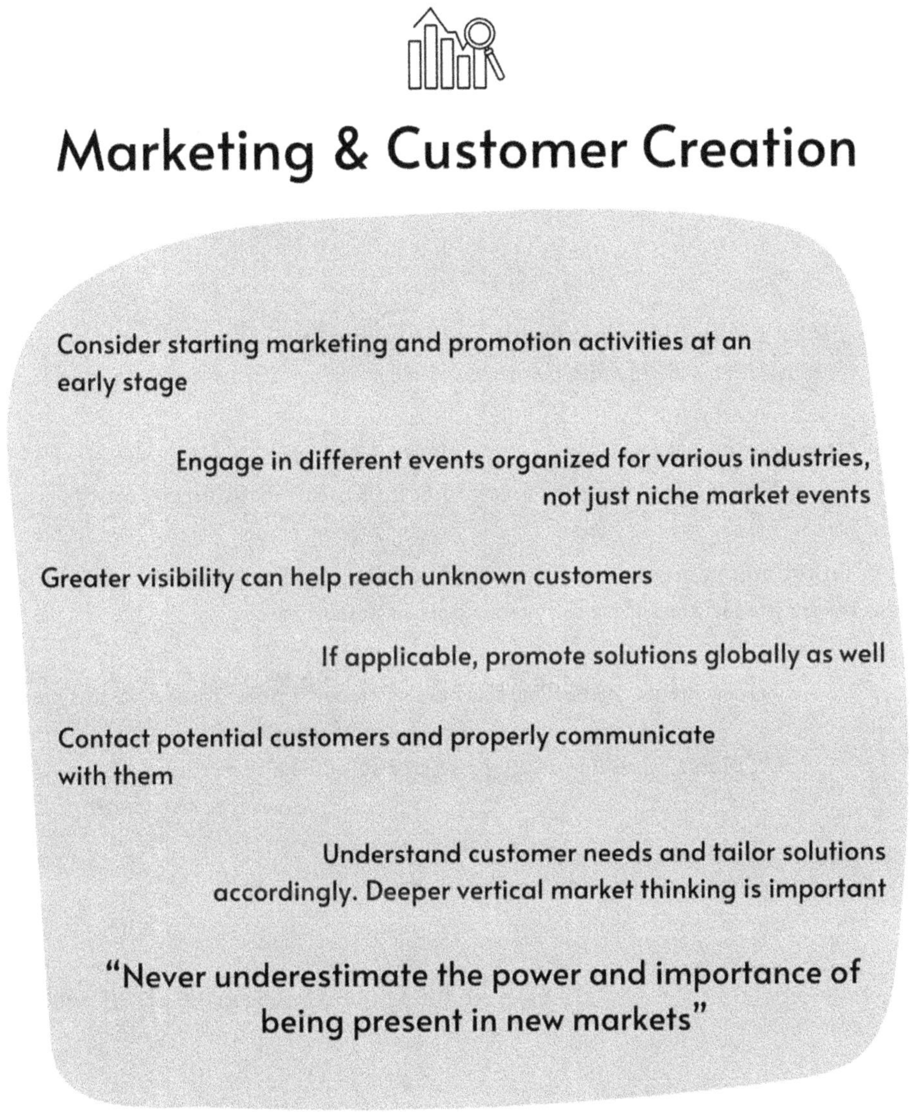

Marketing & Customer Creation

Consider starting marketing and promotion activities at an early stage

Engage in different events organized for various industries, not just niche market events

Greater visibility can help reach unknown customers

If applicable, promote solutions globally as well

Contact potential customers and properly communicate with them

Understand customer needs and tailor solutions accordingly. Deeper vertical market thinking is important

"Never underestimate the power and importance of being present in new markets"

Figure 1.8. Crossing the VoD—some common strategies to find the ideal customers.

of the strategies listed in figure 1.9. A key aspect to remember, especially for startups selling physical products, is to anticipate sales and prepare to meet the sales demand by optimizing production of the product. Nothing brings more displeasure to a customer than not receiving a product within the promised period.

1.5 Tips from a successful academic entrepreneur

Dr Langer has a five-pronged approach to accelerating the pace of academic research and enabling them to become products in the real world. It is as follows:

$$(\$)$$

Sales

Make early sales - early sales efforts help startups to engage
in piloting, determine demand, test customer's commitment

"We see pilots as a way to get foot in the door and proof of concepts
offering a low threshold way to test solution's business benefits"

"Piloting brings in revenues, but we have tried to price them
in a way that makes the main emphasis on learning"

"We offer and bill customers early with pre-study and design
sprint services to (i) teach them about the issue (ii) scope
problems to solve, and (iii) better select the best cases which we
then sell to the customer"

Giving products or services for free does not guarantee
client commitment

Prepare for possible mass production/ sales

Figure 1.9. Crossing the VoD—some common strategies to improve sales.

1. *Choosing high-impact problems:* The first aspect of choosing the right projects, according to Dr Langer, is to choose them based on the societal impact, not the monetary benefits. Customers and money will follow if the project solves a predominant and prevalent problem well. Some of the groundbreaking products from his lab were developed to improve health outcomes for billions of people—for example, a potential cure for Type 1 diabetes where beta cells were encased in a polymer to protect them from the body's immune system at the same time allowing them to detect blood sugar levels and release insulin

appropriately. The second criterion is that the project should align with the core strength of the research facility. The third one is to check if the project's goal can be met by applying or expanding existing scientific concepts at the research facility or in collaboration with other groups.

2. ***Crossing the proverbial 'Valley of Death':*** Dr Langer has a few solutions to help research labs navigate this phase—focusing on solutions that might have multiple applications, obtaining a broad patent, publishing a seminal journal article in a high impact factor esteemed journal, conducting enough preliminary studies (animal studies, if needed) to establish the validity of the discovery before taking steps for commercialization, rewarding the researchers, providing an opportunity of the researchers to be part of the commercial development (either through job roles in the company or as advisors), and ensuring that the companies that license the technology build the company.

3. ***Facilitating multidisciplinary collaboration:*** To develop impactful solutions to the world's greatest healthcare technologies, a key component is a multidisciplinary collaboration between engineering, physics, computational sciences, mathematics, biomedical sciences, and other related experts. Some of the challenges in a multidisciplinary lab are managing diverse expertise, addressing differences that exist in technical languages and experimental techniques, coordinating complex workflows, bridging communication gaps between different disciplines, addressing conflicting priorities, distributing funding, equipment, and other resources fairly, and managing team dynamics and leadership. Therefore, it is important for a research lab to look for people with great team spirit and excellent communication skills, besides having great academic credentials and a passion for research.

4. ***Embracing turnover:*** Dr Langer and many academicians embrace the constant in-and-out movement of human resources as a positive aspect. One main reason is that it puts pressure on the team to achieve results within the finite tenure of the researchers and the limited duration of grants. Most grants do not renew unless goals are met. The other interesting aspect is that a research lab is always composed of people from different generations with varying levels of experience - the wealth of knowledge possessed by a professor and the crazy ideas and foundational questions raised by young researchers are both critical to great scientific research.

5. ***Leadership style that balances freedom and support:*** Researchers thrive on opportunities to explore, where they can learn from mistakes and take innovative risks. Dr Langer exposes his group to possibilities and lets each one choose what they want to work on, thereby, empowering them. In a recent Forbes article (Forbes 2021), he shared that the most successful people, whether in scientific research or business, must learn to be resilient and effectively deal with failure. To enable his students and employees to deal with obstacles and setbacks, which are very common in the entrepreneurial world, he believes that providing positive reinforcement, showing trust and faith, and showing appreciation for their efforts will go a long way toward helping them feel empowered.

1.6 A real-world case study

A fitting case study of how academic research can successfully transition from a research lab into a scalable startup is the story of Eko Health. Eko Health is a digital health company that develops AI-powered stethoscopes and associated analysis software to help healthcare providers detect and monitor heart and lung disease. The company was founded in 2013 by Connor Landgraf, Jason Bellet, and Tyler Crouch.

Ideation: The idea for a smart stethoscope came to Connor while pursuing his master's in mechanical engineering at the University of California, Berkeley, USA. During a course by Professor Amy Herr, who leads the Herr Lab in Bioinstrumentation for Quantitative Biology and Medicine, Connor had the opportunity to connect with clinicians to understand their pain points with technology. Clinicians complained about the difficulty hearing and diagnosing heart conditions using one of medicine's oldest tools—the analog stethoscope. Connon then decided to combine AI and acoustic analysis to improve heart sound detection. He recruited fellow students Crouch, who was a mechanical engineering under-graduate, and Belley, an undergraduate student in the business administration program at Berkeley's Haas School of Business.

IP, patents, and university licensing: The first patent was filed in 2014 (Eko Health Patent 2014). Since the research originated while the team was at UC Berkeley, the formed startup, called Eko Health, negotiated IP rights with the university and got them to license patents in exchange for royalties.

PoC and initial funding: The initial challenges were the need to conduct PoC studies to prove that an AI-enhanced stethoscope could outperform a traditional one and the need for thorough clinical validation. The team first built an early prototype of a smart stethoscope that could digitally capture and record heart sounds, using an initial grant from UC Berkeley's SkyDeck incubator. In 2015, they raised $2.8M in seed funding in a round led by FOUNDER.org and Stanford's StartX. Later in the year, Eko received FDA clearance for its first product, the CORE, a digital attachment for traditional stethoscopes. CORE can be inserted into the tubing of traditional stethoscopes, and it captures the digital recordings of the patient's heartbeats and sends them wirelessly via Bluetooth technology to Eko's HIPAA-compliant smartphone app for analysis. Landgraf, CEO, and his co-founders, Tyler Crouch, CTO, and Jason Bellet, COO, were considered the youngest team to receive FDA clearance for a Class II medical device (Healthcare IT News 2015).

Technology validation: The University of California, San Francisco's Department of Cardiology led its initial trials. They partnered with Stanford University and Mayo Clinic for institutional pilots and validation studies. Since then, Eko Health has been involved in several clinical trials and studies to validate and improve its AI-driven cardiac diagnostic tools.

Patents and papers: Over the years, Eko Health has obtained several hardware, software, and design patents (Eko Health Patents 2025) and published several papers (Eko Health Papers 2025).

Market research and business model: Their market research revealed that over 13 million people globally suffer from undiagnosed heart disease every year. They also decided to take the B2B route to sell their device to healthcare providers, hospitals, and telemedicine providers. They also added a B2B subscription model for using their AI-powered analytics as software as a service (SaaS), ensuring recurring revenue.

Further fundraising: In 2018, Eko Health received $5M Series A funding from ARTIS Ventures, DreamIt Ventures, 1812 Ventures, and FOUNDER.org to grow their commercial team and conduct further clinical studies for using Eko for valvular heart disease screening and heart failure monitoring. In 2019, they closed a $20M Series B round, again led by ARTIS Ventures and others, for further research and development and commercialize Eko's machine learning platform for cardiac screening and analysis. In 2020, they announced $65 million in Series C funding led by Highland Capital Partners and Questa Capital, with participation from ARTIS Ventures and others. This round of funding was raised to expand their market to telemedicine and launch a cardiopulmonary monitoring program. They also announced their partnership with 2M to integrate its digital stethoscope technology with 3M's Littmann stethoscopes. In 2022, they received a Series C extension of $30M. They were also awarded a $2.7M Small Business Innovation Research (SBIR) grant by the National Institutes of Health (NIH) Department of Health and Human Services (HHS) to further their development of a machine learning algorithm that detects and stratifies pulmonary hypertension using phono-cardiogram and electrocardiogram data provided by Eko's smart stethoscopes. Finally, in June 2024, they raised $41M in Series D funding to scale its AI-powered heart and lung disease detection platform.

Further regulatory clearance: In 2017, Eko received another FDA clearance for the DUO, the first combined digital stethoscope and single-lead ECG. In 2019, the company was granted the FDA's Breakthrough Designation for its low ejection fraction screening algorithm, developed in collaboration with the Mayo Clinic, highlighting its potential to address unmet medical needs. According to the FDA (FDA 2025), 'The Breakthrough Devices Program is intended to provide patients and healthcare providers with timely access to medical devices by speeding up development, assessment, and review for premarket approval, 510(k) clearance, and De Novo marketing authorization.' In 2022, the company received 510(k) clearance for its heart murmur analysis software. In 2023, Eko launched the CORE500, a digital stethoscope featuring integrated AI software, high-fidelity audio, a full-color display, and a three-lead ECG.

Key takeaways from Eko health's journey: The above case study proves that academic research can successfully transition into a scalable startup. Some of the key factors that helped Eko Health start and sustain a business are the following:

- Having a strong founding team with diverse technical and business expertise.
- Having early conversations to understand the pain points of clinicians and solving a significantly prevalent problem using cutting-edge technology.
- Securing university IP rights early to speed up commercialization.

- Communicating with regulatory bodies such as the FDA earlier in the development process to enable faster and seamless approval.
- Partnering with top institutions (e.g., Stanford, Mayo Clinic) to conduct early large-scale clinical studies and publish timely patents and journal articles to establish their clinical evidence and validity.
- Innovating continuously to stay ahead of competitors and addressing evolving market needs.
- Securing strategic partnerships with companies (AstraZeneca and 3M) to establish their credibility in the field.
- Having a strong business model (B2B SaaS + hardware) to ensure long-term growth and revenue.
- Focussing on scalability and market expansion by catering to individual clinicians, hospitals, and telemedicine providers.
- Designing solutions that improve health equity to ensure that heart and lung care is accessible to people in low-resource settings globally.

The business model and the path to growth and scaling might not look the same for different companies from different sectors. However, there are always key lessons to learn from other successful companies that can help navigate any existing roadblocks.

1.7 Final remarks

Stepping into entrepreneurship from academia is a bold yet rewarding journey. It is both fulfilling and exhilarating to see an academic invention or discovery in the hands of the real world. Academics hold the immense power to revolutionize healthcare by translating research into real-world solutions. As evident by the journey detailed in this chapter, the transition from an academician to an entrepreneur demands resilience, adaptability, and a willingness to embrace uncertainty, and it also offers an opportunity to move beyond publishing papers and create lasting change. The lessons are worth the end reward. One takeaway from this chapter is that building and partnering with a strong stakeholder network will accelerate and sustain growth and provide the work-life balance necessary to enjoy university and startup responsibilities.

References

ARPA 2025 https://arpa-h.gov/
BARDA 2025 https://aspr.hhs.gov/AboutASPR/ProgramOffices/BARDA/Pages/default.aspx
Blank S 2005 *The Four Steps to the Epiphany: Successful Strategies for Products that Win* 1st edn (Pescadero, CA: K&S Ranch)
Blank S and Dorf B 2013 *The Startup Owner's Manual: The Step-by-Step Guide for Building a Great Company* 3rd edn (Pescadero, CA: K&S Ranch)
Blumberg M 2013 *Startup CEO: A Field Guide to Scaling Up Your Business* 1st edn (New York: Wiley)

Burkholder P and Hulsink W 2022 Academic intrapreneurship for health care innovation: the importance of influence, perception, and time management in knowledge commercialization at a University's Medical Centre *J. Technol. Transf.* https://doi.org/10.1007/s10961-022-09974-6

Eko Health Papers 2025 https://ekohealth.com/blogs/published-research?srsltid=AfmBOopmiF_tRgr2V54gZnMezaL7EKQNozuheLiZD1PvY0HrzY6xL7lV

Eko Health Patent 2014 Mobile Device-Based Stethoscope System (Patent: US9973847B2) https://patents.google.com/patent/US9973847B2/en

Eko Health Patents 2025 https://ekodevices.zendesk.com/h.c./en-us/articles/13156748563227-Patents

Engzell J, Karabag S F and Yström A 2024 Academic intrapreneurs navigating multiple institutional logics: an integrative framework for understanding and supporting intrapreneurship in universities *Technovation* **129** 102892

FDA 2025 Breakthrough Devices Program https://fda.gov/medical-devices/how-study-and-market-your-device/breakthrough-devices-program

Forbes 2021 The modest Moderna cofounder and multibillionaire Robert Langer shares his secrets to success *Forbes* https://forbes.com/sites/jackkelly/2021/11/16/the-modest-moderna-cofounder-and-multibillionaire-robert-langer-shares-his-secrets-to-success/

Gbadegeshin S A, Al Natsheh A, Ghafel K *et al* 2022 Overcoming the valley of death: a new model for high technology startups *Sustain. Fut* **4** 100077

Gooneratne N, McGarrigle R and Winston F (ed) 2021 *Academic Entrepreneurship for Medical and Health Scientists* (PubPub)

Glover W J, Crocker A and Brush C G 2024 Healthcare entrepreneurship: an integrative framework for future research *J. Bus. Ventur. Insights.* **22** e00476

Healthcare IT News 2015 https://healthcareitnews.com/news/fda-clears-new-stethoscope-new-age

Horowitz B 2014 *The Hard Thing about Hard Things: Building a Business When There Are No Easy Answers* 1st edn (New York: Harper Business)

Lim W M, Ciasullo M V, Escobar O and Kumar S 2024 Healthcare entrepreneurship: current trends and future directions *Int. J. Entrep. Behav. Res.* **30** 2130–57

Marcolongo M 2017 *Academic Entrepreneurship: How to Bring Your Scientific Discovery to a Successful Commercial Product* 1st edn (New York: Wiley)

Markham S K 2002 Moving technologies from lab to market *Res. Technol. Manag.* **45** 31–42

Micalo S 2022 *Healthtech Innovation: How Entrepreneurs Can Define and Build the Value of their New Products* 1st edn (New York: Productivity Press)

Meyers A (ed) 2023 *Digital Health Entrepreneurship. Health Informatics* (Cham: Springer)

National Research Council 2000 *From Research to Operations in Weather Satellites and Numerical Weather Prediction: Crossing the Valley of Death* (Washington, DC: The National Academies Press)

POCC 2025 https://seed.nih.gov/programs-for-academics/academic-entrepreneurship-and-product-development-programs

Ries E 2011 *The Lean Startup: How Today's Entrepreneurs Use Continuous Innovation to Create Radically Successful Businesses* 1st edn (New York: Crown Business)

SBIR 2025 https://sbir.gov/

Sieg P, Posadzińska I and Jóźwiak M 2023 Academic entrepreneurship as a source of innovation for sustainable development *Technol. Forecast. Soc. Change* **194** 122695

SPARK 2025 https://sparkmed.stanford.edu/

Shane S 2004 *Academic Entrepreneurship* (Cheltenham: Edward Elgar Publishing)
The Founders Network 2025 https://foundersnetwork.com/blog/startup-financials/
Thiel P and Masters B 2014 *Zero to One: Notes on Start Ups, or How to Build the Future* 1st edn
  (New York: Crown Business)
Upmetrics 2025 https://upmetrics.co/blog/write-financial-section-startup-business-plan

Chapter 2

Can generative AI change how we deliver medical education?

In recent months, the widespread use of large language models (LLMs) such as ChatGPT has sparked several extensive discussions on how they can positively and negatively impact medical education. This chapter explores LLM opportunities in several key areas of medical education, such as curriculum development, teaching methodologies, personalized learning, learning assessment, medical program tracking, and medical research. It also highlights the challenges of model inconsistencies and reliability, privacy, copyright, plagiarism and security, algorithmic bias, limited knowledge, etc, and provides relevant recommendations to address the challenges. The insights gleaned from this chapter would help hospital administrations, educators, and students responsibly utilize this impactful technology for medical education.

2.1 Introduction

We live in an era with unprecedented opportunities to enhance our learning experience via artificial intelligence (AI) and Generative AI technologies. A significant paradigm shift is underway in the field of AI because of the emergence of large-scale self-supervised models called LLMs (Vaswani *et al* 2017, Brown *et al* 2020, Kaplan *et al* 2020) such as OpenAI's Generative Pre-trained Transformers (GPT-4) (OpenAI 2024), Google's Gemini 1.0, PaLM 2, MedLM (GoogleAI 2024), Anthropic's Claude 3.5 (Anthropic 2024), xAI's Grok-1 (xAI 2024), Meta's Llama (Meta AI 2024), etc.

At the most basic level, an LLM is a model capable of receiving input and presenting output in human language. They are called 'large' models because of the large amounts of data they are trained on and the number of parameters involved in the training and deployment process. Over the last couple of years, several applications have come out of the use of these models, especially in tasks related to computer vision (text-to-image, text-to-video models, image captioning, visual question answering), natural language processing (text generation, language translation, summarization, sentiment analysis, question answering), education and training (personalized learning assistants, curriculum development, language education), healthcare (clinical documentation, patient communication via chatbots, research), business (customer support, content marketing, data analysis), research and development (code generation, literature review, scientific discovery), legal and regulatory (compliance checks, contract analysis, legal research), media (content creation, game development, visual storytelling), software development (code generation and explanation, DevOps assistance), etc.

AI, especially LLMs, holds great potential to revolutionize education. However, it is also important to understand that the capabilities of these models are still in the infancy stage, and successful integration into education will be challenging until issues such as algorithmic bias, plagiarism, copyright infringement, privacy, security, misinformation, and hallucinations are addressed.

The objectives of this chapter are to highlight the opportunities and challenges of using AI in the field of education and offer a perspective on its pitfalls. The chapter also highlights future recommendations and best practices to ensure the responsible use of AI in education.

2.2 Opportunities

The integration of AI and generative AI into medical education is quickly impacting traditional approaches to teaching and learning. This section presents the various ways these technologies can potentially impact aspects of education, such as curriculum development, teaching methodologies, etc. Beyond teaching, LLM can also support research, administrative tasks, and program monitoring, positioning itself as a vital enabler of modern medical education. Figure 2.1 summarizes the six key opportunities for using LLMs in medical education.

2.2.1 Curriculum development

The right curriculum embodies various factors, such as aligned lesson plans, the most up-to-date content, measurable learning objectives, and assessment strategies (Davis and Harden 2003). LLMs such as GPT-4 can create detailed lesson plans that form the basis of a roadmap to guide instruction and help the learners achieve their learning goals (Wang *et al* 2023, Lee 2024). The lesson plans can outline the learning objectives, content, and assessment activities, such as rubric cubes and scoring guides. While creating a curriculum mapping and objectives framework, generative AI can convert broad educational goals to specific learning objectives and create measurable and actionable learning outcomes. It can also help map these objectives to align with education standards and timelines (learning path) and set up assessment checkpoints.

Example: Consider a broad educational goal: 'Understand the structure and function of the human heart.' Here is the output from Claude 3.5:

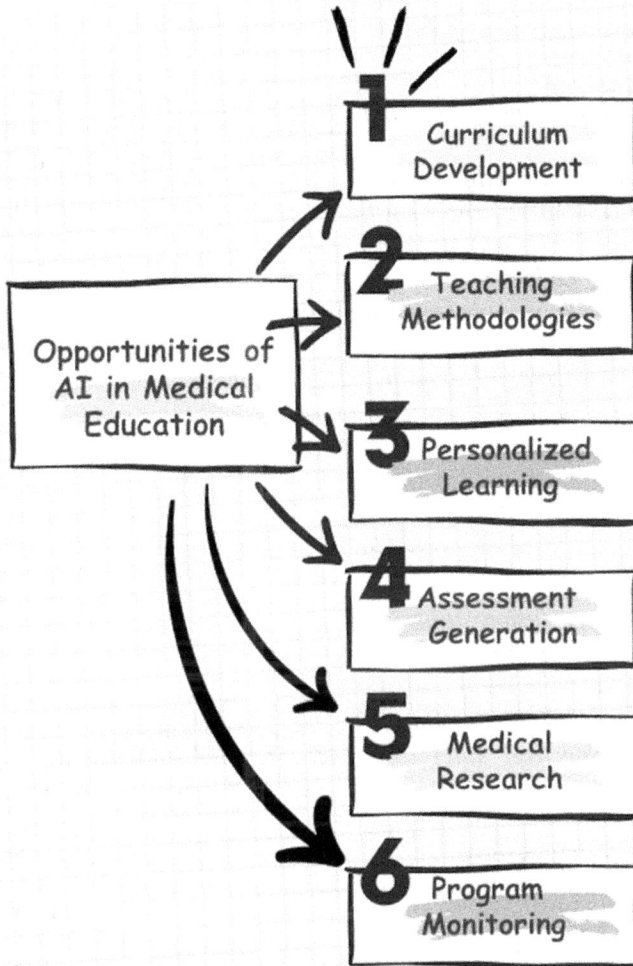

Figure 2.1. Summary of opportunities for using LLMs in medical education.

Level 1: Fundamental Knowledge Objectives

Generated Main Objective: 'Understand the structure and function of the cardiovascular system'

1. **Specific Learning Objectives:**
    1. *Identify major components of the cardiovascular system*

        ○ *List four chambers of the heart*
        ○ *Describe the different types of blood vessels*
        ○ *Identify major blood vessel pathways*
    2. *Explain basic cardiac function*

        ○ *Describe the cardiac cycle*

○ *Define systole and diastole*

○ *Explain blood pressure components*

Level 2: Clinical Application Objectives
Generated Main Objective: *'Apply cardiovascular system knowledge to clinical scenarios'*
1. **Specific Learning Objectives:**
    1. *Interpret cardiovascular measurements*

        ○ *Analyze normal* vs. *abnormal ECG readings*
        ○ *Evaluate blood pressure measurements*
        ○ *Interpret heart sounds and their significance*

    2. *Assess cardiovascular function*

        ○ *Calculate cardiac output*
        ○ *Evaluate perfusion status*
        ○ *Analyze pulse characteristics*

Level 3: Integration and Analysis Objectives
Generated Main Objective: *'Integrate cardiovascular knowledge with clinical decision-making'*

1. **Specific Learning Objectives:**
    1. *Evaluate cardiovascular pathologies*

        ○ *Differentiate between various heart conditions*
        ○ *Analyze risk factors for heart disease*
        ○ *Assess impact of lifestyle on cardiovascular health*

    2. *Develop treatment strategies*

        ○ *Create patient care plans*
        ○ *Select appropriate interventions*
        ○ *Monitor treatment effectiveness*

2. **Alignment to Standards:**
    - *ACGME Patient Care Competency: PC-1 'Gather essential information about patients and their conditions through history-taking'*
    - *USMLE Step 1: Cardiovascular System*
    - *Medical School Objective Project (MSOP): 'Apply knowledge of cardiac physiology to clinical scenarios'*

3. ***Progressive Learning Path:***
   1. *Basic Knowledge (Week 1–2)*

      - *Anatomy and structure*
      - *Basic physiology*
      - *Fundamental concepts*

   2. *Clinical Applications (Week 3–4)*

      - *Physical examination skills*
      - *Diagnostic interpretation*
      - *Basic pathology*

   3. *Integration Phase (Week 5–6)*

      - *Case-based learning*
      - *Clinical problem-solving*
      - *Treatment planning*

4. ***Assessment Checkpoints:***
   1. *Knowledge Verification*

      - *Multiple-choice questions on anatomy*
      - *Labeling exercises*
      - *Concept explanations*

   2. *Skill Demonstration*

      - *Physical examination techniques*
      - *ECG interpretation*
      - *Case analysis*

   3. *Clinical Competency*

      - *Patient case presentations*
      - *Treatment plan development*
      - *Clinical reasoning exercises*

The above output from Claude illustrates the overall learning objectives for the cardiovascular unit for different levels of understanding, from basic to clinical, how these objectives map to current medical education standards, an example of a six-week learning path, and a list of major assessment points. We can continue prompting the LLM to provide lesson plans to meet the above objectives. For example, if we ask for an outline of all lesson plans to align with the above curriculum framework, the LLM responds with the following:

Week 1–2 (Basic Knowledge) lesson plans might include:
- *Heart anatomy and structure*
- *Basic cardiac physiology*
- *Blood flow patterns*
- *Electrical conduction system*

Week 3–4 (Clinical Applications) lesson plans might include:
- *Physical examination techniques*
- *ECG interpretation basics*
- *Common pathology patterns*
- *Clinical investigation methods*

Week 5–6 (Integration) lesson plans might include:
- *Case-based learning sessions*
- *Clinical reasoning workshops*
- *Treatment planning exercises*
- *Patient management scenarios*

If we prompt for one lesson plan focused on physical examination skills specifically, the LLM responds with the below plan:

Lesson Plan: Introduction to Cardiovascular System Examination
Topic: *Physical Examination of the Cardiovascular System*
Duration: *120 min*
Target Audience: *Second-year Medical Students*
Prerequisites: *Basic cardiac anatomy and physiology*
Learning Objectives
By the end of this session, students will be able to:
1. *Demonstrate proper cardiac examination sequence*
2. *Perform accurate pulse examination and blood pressure measurement*
3. *Identify normal heart sounds and their characteristics*
4. *Document cardiovascular examination findings systematically*

Required Resources
- *Simulation lab with examination tables*
- *Stethoscopes*
- *Blood pressure cuffs*
- *Patient simulators*
- *ECG rhythm simulator*
- *Visual aids of normal heart sounds*
- *Documentation templates*

Pre-Class Preparation
Students should review:
- *Basic cardiac anatomy*
- *Normal heart sound characteristics*
- *Pulse points anatomy*
- *Blood pressure measurement technique*

Lesson Structure
1. *Introduction (15 min)*
   - *Brief review of cardiovascular anatomy*
   - *Overview of systematic examination approach*
   - *Demonstration of required equipment*
   - *Safety and patient comfort considerations*

2. *Demonstration Phase (30 min)*
   - *Instructor demonstration of complete cardiac examination*
   - *Step-by-step breakdown of examination sequence:*

     - *Pulse examination*
     - *Blood pressure measurement*
     - *Precordial examination*
     - *Heart sound auscultation*
   - *Real-time commentary on technique and findings*

3. *Guided Practice (45 min)*
   - *Students divided into groups of three*
   - *Rotating roles: Examiner, Patient, Observer*
   - *Practice sequence:*

     - *Pulse examination (10 min)*
     - *Blood pressure measurement (10 min)*
     - *Heart sound auscultation (15 min)*
     - *Complete examination sequence (10 min)*
   - *Instructor circulates providing feedback*

4. *Assessment and Feedback (20 min)*
   - *Brief practical assessment of key skills*
   - *Peer feedback session*
   - *Common errors discussion*
   - *Documentation practice*

5. *Closure (10 min)*
   - *Summary of key points*
   - *Preview of next session*

- *Assignment of practice tasks*
- *Q&A session*

Assessment Methods
- *Direct observation during practice*
- *Peer evaluation forms*
- *Skills checklist completion*
- *Documentation quality review*

Follow-up Activities
- *Practice sessions in skills lab*
- *Online heart sound tutorials*
- *Peer practice groups*
- *Clinical correlation sessions*

Extensions and Modifications
For Advanced Students:
- *Additional pathological heart sound recognition*
- *Complex case scenarios*
- *Teaching junior peers*

For Students Needing Support:
- *Additional guided practice sessions*
- *Simplified examination sequence*
- *Video resources for review*
- *One-on-one tutoring*

Documentation Requirements
Students must submit:
- *Completed examination checklist*
- *Sample examination documentation*
- *Peer evaluation forms*
- *Self-reflection notes*

Safety Considerations
- *Hand hygiene protocols*
- *Equipment sanitization*
- *Professional behavior guidelines*
- *Patient dignity and comfort measures*

As is evident, it is very efficient to generate plan outlines and continue prompting to go to the depth of the details until the required details are flushed out. The depth of details is commendable but must be reviewed and approved by a medical educator for comprehensiveness and correctness. Even still, this technological advancement is enabling educators to quickly provide suggestions on all aspects of the curriculum, freeing up more time for them to focus on other aspects of teaching and guiding students (Dumić-Čule *et al* 2020, Çalışkan *et al* 2022, Ngo *et al* 2022).

Once the lesson plans are ready, the next step is content creation; these high-powered LLMs can easily convert high-level topics to detailed subtopics, creating nuggets at multiple difficulty levels that diverse learners easily comprehend. Besides traditional content, LLMs can help develop lesson supplements quickly (including multimedia content scripts), teaching guides, study guides, teaching aids, reference documents, cross-curricular opportunities, cross-subject lesson plans, and integrated project ideas. They can also be used to develop teaching and study materials in alternative formats and languages, using varied engagement methods and modified assessment tools to improve accessibility and inclusion in education.

Finally, these efficient helpers can also aid the educator in developing effective documentation and communication materials such as program summaries, parent guides, student handbooks, and progress report templates.

2.2.2 Teaching methodologies

Medical education uses several teaching methodologies to empower students to achieve diverse learning objectives, from foundational knowledge to clinical skill development and decision-making. In traditional lecture-based learning, where subject-matter experts teach via lectures, AI can aid in automated content and supplemental material creation and lecture summarization, and chatbots can answer follow-up questions from students. Besides this traditional approach, several other methodologies have found their place in medical education. Some are listed below, along with details on how AI can support these methodologies.

Case-based learning (CBL): Case-based and problem-based learning have shown immense success in helping learners easily attain learning outcomes (Donkin *et al* 2023, Sauder *et al* 2024). A study by Gasim *et al* (2024) determined that the CBL satisfaction rate among medical students at the Faculty of Medicine, National Ribat University, was 92.4%. In CBL, students work through real-life clinical cases and apply their theoretical knowledge to practical contexts. CBL promotes critical thinking and reasoning abilities. AI can be important in generating virtual clinical scenarios to assist in CBL. For example, AI can create diverse, complex patient cases by including realistic patient histories and clinical findings. AI can also integrate rare conditions and atypical presentations, which are hard for medical students to come by in real life.

Problem-based learning (PBL): In the case of problem-based learning, small groups of students collaborate to solve a medical problem. PBL encourages self-directed learning and teamwork. It is generally time-consuming and requires skilled facilitators. To enhance PBL, AI can present complex clinical scenarios and create guided discovery pathways, facilitation guides, and discussion points to support the students' curiosity and critical thinking (Divito *et al* 2024).

Healthcare simulation (HCS): In HCS, students practice clinical skills in simulated environments, using mannequins, virtual patients, or even role-playing. HCS provides a safe setting for students to have hands-on experience with their learned skills while encouraging teamwork and enhancing crisis management skills.

The commonly used tools of HCS are virtual reality (VR), augmented reality (AR), and hi-fi simulators. The recent work published by Hamilton (2024) provides a detailed overview of AI's impact on simulation.

Team-based learning (TBL): In TBL, which is a more structured approach than CBL and PBL, students study the material beforehand, demonstrate understanding via assessments, and then come together as a team to apply the learned knowledge to tackle problems. The goal is to build collaboration and communication skills while ensuring accountability for each individual's role. AI can facilitate group discussions, assess team and individual dynamics, and create customized preparatory material for each team.

Flipped classroom: To allow students to shift from passive to active learning, the flipped classroom methodology assigns review materials to students before class and allows them to engage in interactive activities during class. AI can be powerful in helping educators develop personalized pre-class learning paths based on each student's capabilities.

Clinical rotations and bedside teaching: Clinical rotations based on hands-on learning are a hallmark of medical education, where students are directly exposed to patient care to aid them in bridging theoretically learned knowledge with real-world applications. When clinical rotations are difficult, AI-powered virtual patients can offer practice for clinical reasoning. AI tools like AR overlays can help students interpret lab results or imaging on rounds. AI can also help draft clinical notes during the rounds.

E-learning and digital platforms: These include online courses (Coursera, Osmosis, etc), mobile apps, interactive modules, etc, that can deliver education via flexible and scalable options. It is easy for such digital platforms to incorporate customized content delivery based on learner behavior and progress.

Competency-based medical education (CBME) focuses on helping students achieve specific competencies rather than time-based outcomes. This methodology ensures that students are ready for independent practice. Since this methodology is more personalized, AI can play a significant role in assessing and tracking individual progress toward competency milestones and providing recommendations.

Peer-assisted learning (PAL): As the name indicates, in PAL, students teach and learn from their peers. AI can help student tutors in creating customized teaching aids.

Evidence-based medicine training (EBM) emphasizes integrating clinical expertise with knowledge available in research articles. AI is already very prevalent in providing research-backed information (Perplexity AI) and has demonstrated a good ability to extract and summarize key points from research papers. It can also summarize the latest evidence and propose clinical applications. Figure 2.2 shows the response of Claude 3.5 Sonnet model (Perplexity AI) to the prompt 'Summarize the latest evidence around CRISPR gene editing technology and propose its clinical applications.' It also includes links to references for further reading.

Gamification is a methodology that uses game-based techniques to teach medical concepts. It is an effective and engaging way to help students to retain information. In health, serious games can be used for diagnostics, therapy, and prevention, as well as for medical or patient education (Wang *et al* 2022). For example, in 2020, the US

Let me analyze the breakthrough in CRISPR gene editing technology and its clinical applications:

Clinical Evidence for CRISPR Therapy

CRISPR technology achieved significant clinical validation in 2024 with two major breakthroughs:

FDA-Approved Clinical Trials
Intellia Therapeutics received FDA clearance for a pivotal phase 3 trial treating hATTR, demonstrating CRISPR's potential as a therapeutic platform ⁵. The same Cas9 mRNA system was effective for treating different diseases by modifying just 20 nucleotides of guide RNA ⁵.

First Regulatory Approval
CASGEVY™, developed by Vertex and CRISPR Therapeutics, became the first CRISPR/Cas9 gene-edited therapy approved by the UK MHRA for treating:

- Sickle cell disease
- Transfusion-dependent beta-thalassemia ⁵

Clinical Applications

Diagnostic Tools
The CRISPR-Cas system shows promise for:

- Early cancer detection using CRISPR-Cas12a and CRISPR-Cas13a biosensing
- Precise diagnosis of early-stage cancers
- Cancer therapy monitoring ⁵

Therapeutic Uses

- Treatment of genetic blood disorders
- Personalized genetic modifications
- Platform technology for multiple diseases using the same delivery system with modified guide RNA ⁵

This breakthrough represents a significant advancement in genetic medicine, offering new possibilities for treating previously challenging genetic conditions through precise genome editing.

⊕ Ask follow-up   ● Pro ↑

Figure 2.2. An example of Perplexity AI output on CRISPR technology and its clinical applications, including references. Image Credit: Perplexity AI (2024).

Food and Drug Administration permitted the marketing of EndeavorRx as the first game-based digital therapeutic device to improve attention function in children with attention-deficit/hyperactivity disorder (ADHD) (Kollins *et al* 2020). Another game called EMERGE is a simulation game that presents an emergency department in real-time to aid medical students and physicians in improving their clinical reasoning skills (Chon *et al* 2019). Airway Ex (surgical airway simulation) trains anesthesiologists and emergency physicians in airway management using AI-enhanced scenarios (Yau *et al* 2021).

Reflective Practice: Medical students spend a significant amount of time (on an average of 1.5–6 h per day) on reading (Klatt and Klatt 2011). Even though they are so used to traditional teacher-centered lectures, it is important to use active learning and teaching methodologies outlined above to meet learning outcomes efficiently. Since medical education relies on memory and retention, adopting evidence-based reflective practice strategies to enhance learning efficiency, such as active recall, memory cues, and visual mnemonics, is essential (Arango-Ibanez *et al* 2024). LLMs like ChatGPT can act like virtual coaches to aid students in this methodology. An example of an active recall approach is shown in figure 2.3. Memory cues are strategies where one transforms information in a way that makes it easier to recall. For example, students commonly used the acronym TRAP to remember the symptoms of Parkinson's disease. TRAP stands for Tremor, Rigidity, Akinesia/Bradykinesia, Postural instability. LLMs like ChatGPT can easily come up with such mnemonic or short stories or poems to help students remember information. A visual mnemonic is where visual imagery is used to improve information recall. DALL·E 3, the AI system created by OpenAI, can generate images based on text prompts and can be used for the creation of visual mnemonics.

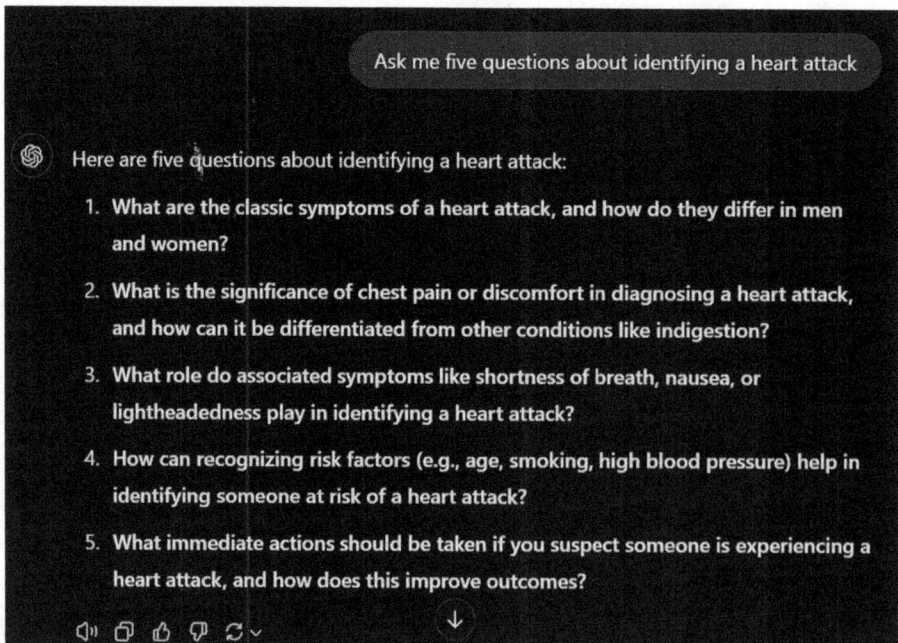

Ask me five questions about identifying a heart attack

Here are five questions about identifying a heart attack:

1. What are the classic symptoms of a heart attack, and how do they differ in men and women?

2. What is the significance of chest pain or discomfort in diagnosing a heart attack, and how can it be differentiated from other conditions like indigestion?

3. What role do associated symptoms like shortness of breath, nausea, or lightheadedness play in identifying a heart attack?

4. How can recognizing risk factors (e.g., age, smoking, high blood pressure) help in identifying someone at risk of a heart attack?

5. What immediate actions should be taken if you suspect someone is experiencing a heart attack, and how does this improve outcomes?

Figure 2.3. An example of ChatGPT output for an active recall approach to learning. Image Credit: OpenAI (2024). ChatGPT Free Version (May 2024 release), [GPT-4o model].

Narrative medicine: This methodology is used to help medical students gain a holistic understanding of patient experiences. It uses storytelling and reflective writing to improve communication skills and foster empathy. AI can help evaluate the written narratives and suggest reflective writing improvement. It can also help create patient stories that students can use in empathy exercises.

Interprofessional education: Human bodies should be treated as a whole, and so collaborative learning among students from different healthcare specialties is essential. This methodology prepares students for team-based care delivery, and AI can create interprofessional teamwork scenarios and role-playing exercises, evaluate team progress, and provide feedback.

2.2.3 Personalized learning

Rabie (2023) conducted a systematic review of the existing literature on the role of AI and personalized education in medical curricula. The review includes 20 studies published between 2013 and 2023, focusing on applications such as intelligent tutoring systems, personalized feedback, and data analysis for individualized instruction. To enable personalized learning, it is important to analyze student performance patterns, learning preferences, knowledge gaps, and accessibility to learning methodologies in real-time. Let us say a student is excellent at physical examination techniques but struggles with interpreting electrocardiogram (ECG) readings. An AI system that recognizes this can easily customize the learning path to include additional ECG practice cases while maintaining a standard pace in other areas the student is comfortable in. Thus, AI can create a dynamic learning environment that improves students' progress. AI can also identify the learning preferences of students through pattern analysis and modify content delivery accordingly. For example, one student may better understand concepts via visual representations (e.g., more animated simulations of heart function). In contrast, another could prefer textual material or hands-on practice (e.g., written descriptions with accompanying case studies). The AI system can also generate customized practice questions to keep the students moving along their optimal learning path.

The same is the case with clinical skills development. AI can create personalized patient scenarios based on the student's level of understanding. For example, if a student struggles with the diagnosis of respiratory conditions, the system could generate more cases with subtle variations in breathing patterns until the student's correctness and confidence grow. Then, it can progressively increase the complexity of the cases so that the student can continue to learn without becoming overwhelmed.

The icing on the cake in AI personalization is in its predictive capabilities. It can monitor learning behavior and performance and predict when a student might struggle with concepts. It can then proactively alter the content delivery materials and methodology to adapt to the student's needs. For instance, if a student's performance pattern suggests they might have difficulty with an upcoming nephrology module based on their previous struggles with related concepts, the system can automatically generate preparatory materials tailored to their specific needs before they encounter difficulties.

2.2.4 Assessment generation

AI can evaluate the impact of various teaching methodologies, as highlighted in the previous section. It can also assess theoretical knowledge via optical mark recognition (OMR), automated essay scoring, and quizzes. It can also help generate diverse question types, rubrics and scoring guides, formative assessments, and feedback templates to aid in assessments. Recently, the National Board of Medical Examiners said it has successfully implemented AI-assisted item writing, with approximately 55% of AI-generated questions meeting expert reviewer approval standards (NBME 2024). Virtual reality can be used to assess the procedural skills of a student by presenting simulated patients in a virtual environment (Pottle 2019). Without VR, simulation assessment can be done by AI, which compares two procedural videos, one by the instructor and one by the student, and provides feedback.

Medication errors are common in the healthcare system. In a study that analyzed over 6700 charts, it was reported that the majority of medical errors are transcription errors followed by prescription and administration errors (Zirpe *et al* 2020). Though AI can also be used to be more vigilant in verifying medication charts, training medical students on proper prescription writing skills is critical. AI can generate incorrect prescriptions as learning material and ask the students to evaluate the prescription for mistakes. AI can also assist students in writing quality clinical documentation by generating sample notes and providing feedback on writing clarity. Responding to medical emergencies is an important skill for a medical professional. AI can be trained with preloaded data on the management of different emergency conditions and be used to improve the competencies of students in emergency response.

2.2.5 Medical research

Grant writing is a time-consuming and labor-intensive process. AI can significantly support, enhance, and speed up the medical research grant writing process. Creating a structured grant proposal outline is the hardest part, and this is where LLMs can be of great use. They can easily and quickly generate structured outlines, methodology descriptions, and literature reviews that funding agencies require. They can also help write compelling research significance statements and even justify budget requirements by presenting the need for resources and personnel. For example, Godwin *et al* (2024) recently presented a software application for writing a typical NIH-style grant. It has three components: (1) a search and compare tool that compares the proposed research to other previously funded NIH applications, (2) a specific aims page drafter, and (3) a research strategy section.

We live in an age of information overload, where the rapid pace of research makes it challenging to stay current. LLMs have intelligent document processing capabilities and can provide immense relief in helping students grasp the core concepts quickly. For example, Petal AI (Petal 2024) is an AI-powered document analysis platform that allows users to search, organize, and get insights from uploaded documents. When students upload research papers, Petal can automatically highlight key findings, help them understand the context and meanings of words and sentences, generate summaries, and create structured notes and practice questions related to the paper. While writing research papers, students can also use LLMs to quickly generate the paper's outline. LLMs can also suggest ways to present, discuss, and analyze the results.

2.2.6 Program monitoring and other admin support

A curriculum management system is a digital platform that helps teaching institutions manage and improve their curriculum. It generally collects data and meta-data related to student achievement, learning outcomes, assessment results, course progress, teacher feedback, student demographics, curriculum alignment with standards, instructional materials used, student engagement levels, and feedback from stakeholders like parents and administrators. LLMs trained on these data can help improve and update the curriculum based on the needs and lack observed. AI can also help draft PowerPoint presentations, letters to patients, and appointment reminders and address other communication-related needs.

2.2.7 Other applications

Recently, OpenAI announced the public release of its much-awaited text-to-video model, Sora (Japanese for 'sky' meaning 'limitless creative potential'). Figure 2.4 depicts some questions on how this model could be utilized to enhance medical education and research. The current video length is limited to 20 s. Prompting clinically accurate medical videos will take significant practice.

2.3 Challenges and recommendations

In medicine, more than in any other field, accuracy, interpretability, and liability are critical, as the outcome directly impacts human lives. Over the past two years, several researchers have tested ChatGPT and other LLMs using various medical licensing exams to verify if they possess accurate knowledge. Liu *et al* (2024) reviewed the literature to understand ChatGPT's performance in such exams. They concluded that though GPT-4 showed immense potential for future use in medical education, it still has several issues that need to be addressed. This section highlights many of those issues in detail and presents some recommendations to address/ mitigate these issues. Figure 2.5 depicts the most common challenges in using AI and Generative AI in medical education.

2.3.1 Lack of reliability

'Hallucination' in AI is a term that denotes that the model has produced seemingly factual results that are either false or misleading. Faithfulness hallucinations, where AI-generated outcomes are divergent from the context provided by the user, are detrimental to high-liability fields such as healthcare. Such responses indicate a lack of reliability, and if students do not scrutinize the validity of the information and end up blindly believing, it could lead to misdiagnoses and inappropriate treatments

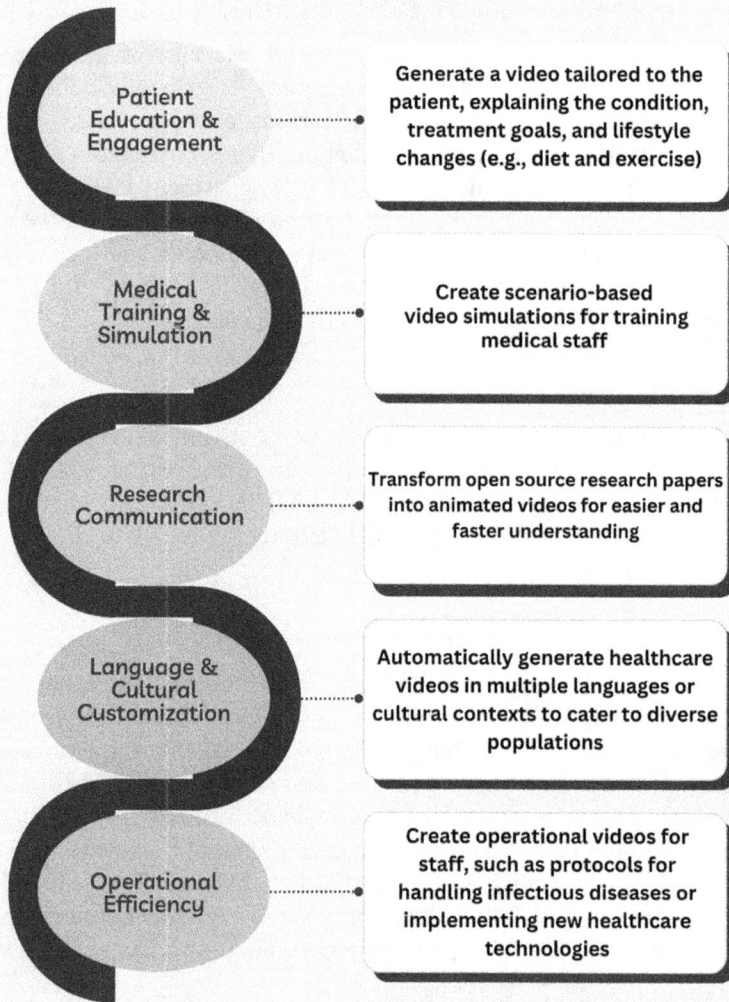

Figure 2.4. OpenAI's text-to-video model Sora's possible uses in medical education.

(Bair and Norden 2023), increasing their malpractice risk (Mello and Guha 2023). There have been reports of citations being fabricated (Goddard 2023). Recent news is that OpenAI's model Whisper frequently invents text passages when presented

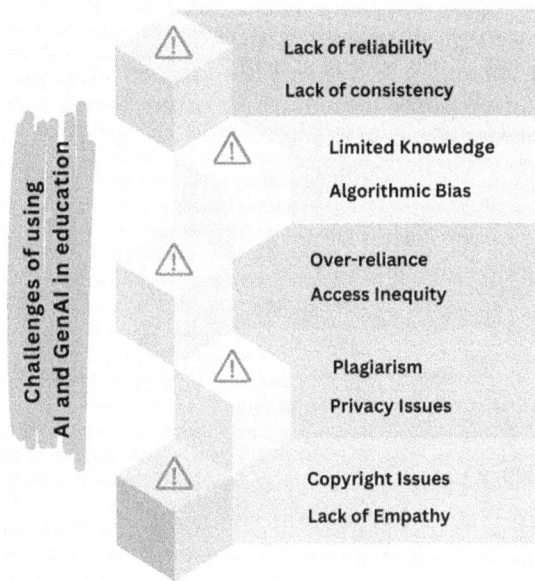

Figure 2.5. Common challenges using AI and GenAI in medical education.

with silence' to say instead: Recent news is that OpenAI's model Whisper frequently invents text passages when presented with silence (Verge 2024).

Hallucinations are not simply errors that can be avoided using better training data, more advanced architectures, or more robust outcome-checking procedures. This is because of several reasons. Firstly, irrespective of how vast the training data is, it will be only an incomplete representation of the total available knowledge. Secondly, LLMs are based on probabilistic principles, so there is always an inherent chance for uncertainties. Thirdly, LLMs do not possess the reasoning capabilities of humans, so they are bound to present seemingly possible but entirely incorrect results. Finally, LLMs operate within a fixed context window; if the required information does not fall within that context, it can lead to hallucinations. How can we address this issue then? We have to accept that hallucinations cannot be completely eliminated and will remain an inherent feature of LLMs. Below are some ways to handle them:

- *Education and awareness:* It is important for students and teaching faculty to be aware of this issue and exercise extreme caution when evaluating the medical information presented by these LLMs. Legally speaking, physicians and students should use these technologies only to augment and not to replace their judgment (Haupt and Marks 2023).
- *Implement verification processes:* Users should verify the generated output by cross-referencing it with authoritative sources or using multiple AI models to corroborate information.

- *Domain-specific fine-tuning:* Fine-tuning models using knowledge from specific domains could potentially reduce domain-specific false information.
- *Prompt engineering:* A prompt is the user input given to LLMs. Prompt engineering is emerging as a new field of research that studies how to design and implement appropriate prompts to ensure the desired output in various domains. Training educators and students to prompt correctly could also potentially reduce hallucinations and false information (Meskó 2023, Wang *et al* 2024).
- *Continued research:* As research in this field evolves, there might be more ways to mitigate hallucinations.

Obviously, human-AI collaboration would still be needed to ensure that AI augments human intelligence and that humans provide critical thinking to complement the AI's capabilities.

2.3.2 Lack of consistency

It has been observed that LLMs sometimes generate different results for the same prompt input. Just like hallucinations, such a lack of consistency in responses makes it harder for students/educators/users to trust the system, especially when the same prompts produce contradicting responses or responses of varying quality (Tlili *et al* 2023). The consistency of generated results could also be improved by using effective prompts. Wang *et al* (2024) used the same set of prompts in different LLMs and asked osteoarthritis-related questions to study the effectiveness of prompt engineering. It was observed that a combination of gpt-4-Web and a prompting technique called reflection on search trees (RoT) presented the best LLM treatment recommendation closest to the clinical guidelines. RoT is an LLM reflection framework designed to improve the performance of tree-search-based prompting methods (Hui and Tu 2024). The other aspect is about how we assess the clinical knowledge of models. Many rely on automated evaluations based on limited benchmarks. Singhal *et al* (2023) recently developed MultiMedQA, a benchmark combining six existing medical question-answering datasets, and proposed a human evaluation framework to assess the model answers along multiple axes, including factuality, comprehension, reasoning, possible harm, and bias. Future research should aim to develop benchmark and human evaluation frameworks involving medical professionals and patients.

2.3.3 Limited knowledge

LLMs come with a knowledge cutoff date, when the training data was last updated. Beyond this date, the model does not know events or information. LLMs can access real-time information through a separate controlling layer to perform web-search and retrieve answers relevant to the user query. However, they cannot use this latest real-time information to train themselves as new information emerges, resulting in incomplete and sometimes out-of-date results. The field of medicine is evolving rapidly, with new research data and findings, guidelines, new digital health tool approvals by regulatory bodies, new treatment protocols, etc. Until LLMs can consistently keep up with newer information, they will fail to provide thorough and latest guidance to healthcare professionals and students. Here are a few ways to keep LLM's knowledge base up-to-date: fine-tuning on domain-specific smaller, targeted datasets, prompt engineering to ensure that specific context is provided, implementing retrieval-augmented generation (RAG) where the LLMs can retrieve relevant information from external sources or databases, and regularly retraining the model on new data.

2.3.4 Algorithmic bias

As the old saying goes, 'Garbage in, garbage out.' That is true in the case of any AI training, including LLMs. Though LLMs are trained on a vast amount of data, if most of the training data for a specific disease comes from a certain ethnic group, then the LLM's responses would be based on that group. Though the generative models are evolving rapidly in addressing this issue, there have been reports that they still perpetuate race-based medicine in their responses (Omiye *et al* 2023). An interesting analysis by Yang *et al* (2024) found that these models tend to associate higher costs and longer hospitalizations for the White populations. They also associate higher survival rates with this population when facing challenging medical scenarios. They also observed biases in generated patient backgrounds, treatment recommendations, associating specific diseases with certain races, etc. Therefore, LLMs must be trained with diverse and representative training data and actively monitored via thorough testing and evaluation to reduce and address bias in the generated results. Fairness metrics can also quantify bias, and adversarial testing can be used to intentionally check the model with biased prompts. If the training data is inherently biased, debiasing algorithms could address it (Yogarajan *et al* 2024). Again, prompt engineering is key, as carefully designed prompts can generate neutral and unbiased responses.

2.3.5 Over-reliance

Another prevalent concern, especially in medical research and student coursework, is the risk of students' overreliance on such technologies to support their research and coursework (Abd-Alrazaq *et al* 2023, Dergaa *et al* 2023). This habit can be an issue, not just from the point of misuse of LLMs, but more importantly, from the perspective of eroding their scholarly abilities and critical thinking abilities. In this era of instant gratification, such technologies can be very tempting for students, leading to decreased motivation to conduct independent research. To help students cultivate critical thinking and prevent them from over-relying on such technologies, educators have to act as 'learning navigators' (Li *et al* 2024). Besides guiding students on correctly using LLMs for academic purposes, educators should pose open-ended questions to encourage critical thinking and facilitate group discussions, debates, role-playing, etc.

2.3.6 Access inequity

Even though generative AI tools are now better at communicating in several languages, proficiency in each language depends on the amount and quality of training data used in that language. Since most training data is in English, LLMs naturally become more accessible and accurate for the English-speaking population. Other key inequities lie in cost ($20–$200 for GPT-4), AI literacy, technology availability, and disabilities (blind, etc). Since internet use is mostly widespread worldwide, some ways to improve access is to establish international university and college collaborations to provide AI-based medical education programs. Open-access online courses and platforms such as Coursera also contribute to improved access for students in remote areas. Specific programs that collect data from non-English-speaking regions and training domain-specific/region-specific models could be another solution. Regarding access to disabled students, AI is great at using features such as text-to-speech, screen readers, voice commands, and adjustable font sizes to accommodate students with visual impairments. Care should be taken to train educators to implement AI strategies to support students with disabilities effectively.

2.3.7 Plagiarism

Academic dishonesty is a known and expected byproduct of such technologies (Dergaa *et al* 2023). Even though there are tools such as Turnitin AI writing detector, Grammarly detector, etc, it is still possible to make AI-generated medical essays undetectable (Tlili *et al* 2023, Scarfe *et al* 2024). Here is where educators should play a key role as 'learning navigators,' as described in section 2.3.5.

2.3.8 Privacy issues

As with any modern-day technology, privacy and security have become inherent issues. Students and educators might inadvertently reveal personal identifying information. This becomes a privacy issue, as OpenAI acknowledges that it may use such information for several purposes, including sharing it with third parties without explicit user consent (Markovski 2023). To address these issues, institutions should implement robust data anonymization techniques by removing patient health information such as names, medical record numbers, dates of birth, etc. When domain-specific LLMs are built, the developers could add noise to sensitive data during training (differential privacy). Furthermore, the first important aspect is only using necessary patient data for training or prompting. Institutions should develop clear guidelines and policies to outline how to use LLMs in medical education. Both educators and students should be trained in the responsible use of these AI tools. It is also important to choose LLM providers with strong data privacy and security practices and compliance with healthcare regulations such as HIPAA and GDPR.

2.3.9 Copyright issues

Gervais *et al* (2024) recently published an article with insights on navigating copyright challenges posed by LLMs. They discuss how an LLM can violate copyright in two ways: (1) using copyright-protected material in the training datasets that can result in the creation of unauthorized copies during the training phase and unauthorized copies of representations of training data embedded within the LLM; and (2) because of the generated results which may be the same or similar to copyrighted material. They also highlight how the international landscape around these issues is inconsistent and is subject to court decisions in many ongoing cases. One recommendation is that global licenses (both direct and collective) can reduce uncertainty and allow copyright owners and LLM users to understand how to use copyrighted works. The other issue is authorship rights for articles written using these generative tools and how students should acknowledge the use of these tools in their work (Abd-Alrazaq *et al* 2023, Dergaa *et al* 2023).

2.3.10 Lack of empathy

One of the key aspects of medical education is the importance of teaching empathy to the students. Fostering empathy is not a skill that LLMs can cultivate. Though it might seem that these tools could be trained to express empathy, they cannot be trained to show genuine emotional experiences. Sorin *et al* (2024) recently conducted a study to review the literature on the capacity of LLMs to show empathy. They concluded that although LLMs exhibit some elements of cognitive empathy by providing emotionally supportive responses, a lot needs to be improved in this area. Another difficult area to address is their inability to perceive and convey non-verbal emotional cues. Students generally learn to develop empathy by observing inter-personal relationship experiences that teaching physicians show their patients. However, LLMs are incapable of perceiving and conveying non-verbal emotional cues. For example, during a ChatGPT-aided simulation at the Yale Center for Healthcare Simulation, a woman presented with abdominal pain, and her concerned boyfriend was at her side. The team that studied this simulation deduced that the woman might have an ectopic pregnancy but did not ensure that the boyfriend left the room prior to them discussing this sensitive issue with the patient. AI-based tools still do not consider such interpersonal dynamics (Safranek *et al* 2023).

2.4 Conclusions and future directions

In the recent American Boards of Medical Specialties Conference 2024, plenary speakers spoke about the transformative nature of AI in medical education. One of the speakers even shared 'AI won't replace doctors, but doctors who know how to use AI will replace doctors who don't. There may come a day when a doctor's proficiency with using AI for practice will need to be assessed.' Such powerful statements indicate the importance of empowering physicians and medical students with the power of AI.

This chapter presented an overview of the opportunities and challenges of using generative AI tools such as ChatGPT in medical education. It seems that when used responsibly, these tools can give back time to the already burnt-out teaching physicians. Those using these tools might be better positioned and high-performing than those who do not. Whether these technologies may take over the menial tasks of humans or not is a question that can be pondered over later. However, first, as the challenges indicate, one should proceed with caution while using these tools, and guardrails should be in place to prevent misuse.

Below are some future directions for responsibly using these tools in medical education.

- **Provide adequate training:** As these tools become more commonplace, academic institutions should take the responsibility to strategize and develop guidelines or best practices quickly. There should be guidance on using, citing, and disclosing content generated by LLMs in research, assignments, etc. They should ensure that such content is detected by software such as Turnitin, Grammarly, Originality.AI, etc, provided these tools are first improved for detecting such content more accurately. It is also upon the institutions to conduct training sessions to empower their educators and students about the ethical use of such tools and to stress the importance of teaching and learning methodology that has components of both human and AI aspects.

- **Update medical curricula:** Current medical curricula do not include much content on using AI for teaching and learning purposes and how AI is used in healthcare delivery. First, educators should understand that AI is here to stay and upskill their competencies in utilizing these tools effectively, as explained in the opportunities section of this chapter. Second, physicians should also upskill themselves on using AI-based digital health technologies for preventative health and disease diagnosis, management, and treatment. They could also aim to be part of the development of such technologies, as no one knows more about what they need in healthcare delivery than the physicians themselves. This would also enable them to develop trust in such technologies and confidently teach their students. Finally, the medical curriculum should be updated with all these elements of AI.

- **Rethink assignments:** Rather than traditional assignments, educators should design assignments that encourage students to think creatively and collaboratively so that the human element is still part of the process. Some examples include oral exams, presentations, hands-on activities, group projects, and case studies. Some assignments could also encourage students to check, critique, and develop better prompting skills to ensure the accuracy of the generated LLM responses.

- **Expert-review content, always:** It might seem addictive to ask these powerful tools to generate lesson plans, study plans, and material and call it a day. However, as highlighted throughout this chapter, LLMs are still learning and susceptible to various mistakes. Therefore, it is important to validate the content using subject-matter experts.

- **Collaborative development:** A collaborative effort is required from educators, students, academic institutions, researchers, and developers of generative AI tools and LLMs for several obvious reasons. However, the two key reasons are (1) to increase the confidence level of educators in the ability of such technologies and (2) to ensure that expert-validated content goes into the training of these LLMs.

- **Research:** Aster *et al* (2024) reviewed several scientific publications on the use of LLMs in medical education. They observed that most studies assessed LLMs' (especially ChatGPT's) performance in medical exams and identified a research gap where these studies lacked an empirical approach and rigorous study design. Therefore, more studies with stringent research designs are needed to evaluate LLMs for accuracy and study the impact of LLM use in medical education.

Overall, generative AI tools haves the potential to impact the medical school curriculum, learning, and teaching methodologies. Just like how we no longer think about electrons when we switch on a light or think about HTTP protocols when we browse, AI will weave into everything we do and become ubiquitous. So, educators and students should be encouraged to embrace such technologies. With collaborative effort among users and developers, it is possible to responsibly use these technologies in a positive light. Now, going back to the question posed as this chapter's title, *'Can Generative AI change how we deliver medical education?'* The simple answer is *'Yes, but proceed with caution'*.

References

Abd-Alrazaq A, AlSaad R, Alhuwail D *et al* 2023 Large language models in medical education: opportunities, challenges, and future directions *JMIR Med. Educ.* **9** e48291

Anthropic 2024 https://anthropic.com/news/claude-3-5-sonnet

Arango-Ibanez J P, Posso-Nuñez J A, Díaz-Solórzano J P and Cruz-Suárez G 2024 Evidence-based learning strategies in medicine using AI *JMIR Med. Educ* **10** e54507

Aster A, Laupichler M C, Rockwell-Kollmann T *et al* 2024 ChatGPT and other large language models in medical education—scoping literature review *Med. Sci. Educ.* **35** 555–67

Bair H and Norden J 2023 Large language models and their implications on medical education *Acad. Med.* **98** 869–70

Brown T B, Mann B, Ryder N *et al* 2020 Language models are few-shot learners https://arxiv.org/abs/2005.14165

Çalışkan S A, Demir K and Karaca O 2022 Artificial intelligence in medical education curriculum: an e-Delphi study for competencies *PLoS One* **17** e0271872

Chon S H, Timmermann F, Dratsch T *et al* 2019 Serious games in surgical medical education: a virtual emergency department as a tool for teaching clinical reasoning to medical students *JMIR Serious Games* **7** e13028

Davis M H and Harden R M 2003 Planning and implementing an undergraduate medical curriculum: the lessons learned *Med. Teach.* **25** 596–608

Dergaa I, Chamari K, Zmijewski P and Ben Saad H 2023 From human writing to artificial intelligence generated text: examining the prospects and potential threats of ChatGPT in academic writing *Biol. Sport* **40** 615–22

Divito C B, Katchikian B M, Gruenwald J E and Burgoon J M 2024 The tools of the future are the challenges of today: the use of ChatGPT in problem-based learning medical education *Med. Teach.* **46** 320–2

Donkin R, Yule H and Fyfe T 2023 Online case-based learning in medical education: a scoping review *BMC Med. Educ.* **23** 564

Dumić-Čule I, Orešković T, Brkljačić B, Kujundžić Tiljak M and Orešković S 2020 The importance of introducing artificial intelligence to the medical curriculum—assessing practitioners' perspectives *Croat. Med. J.* **61** 457–64

Gasim M S, Ibrahim M H, Abushama W A, Hamed I M and Ali I A 2024 Medical students' perceptions towards implementing case-based learning in the clinical teaching and clerkship training *BMC Med. Educ.* **24** 200

Gervais D J, Shemtov N, Marmanis H and Zaller Rowland C 2024 The heart of the matter: copyright, AI Training, and LLMs *SSRN* http://dx.doi.org/10.2139/ssrn.4963711

Goddard J 2023 Hallucinations in ChatGPT: a cautionary tale for biomedical researchers *Am. J. Med* **136** 1059–60

Godwin R C, DeBerry J J, Wagener B M, Berkowitz D E and Melvin R L 2024 Grant drafting support with guided generative AI software *SoftwareX* **27** 101784

GoogleAI 2024 https://ai.google/get-started/our-models/

Hamilton A 2024 Artificial intelligence and healthcare simulation: the shifting landscape of medical education *Cureus* **16** e59747

Haupt C E and Marks M 2023 AI-generated medical advice-GPT and beyond *JAMA* **329** 1349–50

Hui W and Tu K 2024 RoT: enhancing large language models with reflection on search trees arXiv:2404.05449 https://arxiv.org/abs/2404.05449

Kaplan J, McCandlish S, Henighan T *et al* 2020 Scaling laws for neural language models ArXiv:2001.08361 https://arxiv.org/abs/2001.08361

Kollins S H, DeLoss D J, Cañadas E *et al* 2020 A novel digital intervention for actively reducing severity of paediatric ADHD (STARS-ADHD): a randomised controlled trial *Lancet Digit. Health* **2** e168–78

Klatt E C and Klatt C A 2011 How much is too much reading for medical students? Assigned reading and reading rates at one medical school *Acad. Med.* **86** 1079–83

Lee H 2024 The rise of ChatGPT: exploring its potential in medical education [published correction appears] *Anat. Sci. Educ.* **17** 1779

Li Z, Li F, Fu Q, Wang X, Liu H, Zhao Y and Ren W 2024 Large language models and medical education: a paradigm shift in educator roles *Smart Learn. Environ.* **11** 26

Liu M, Okuhara T, Chang X *et al* 2024 Performance of ChatGPT across different versions in medical licensing examinations worldwide: systematic review and meta-analysis *J. Med. Internet Res.* **26** e60807

Markovski Y 2023 Data usage for consumer services FAQ *OpenAI* https://help.openai.com/en/articles/7039943-data-usage-for-consumer-services-faq

Meskó B 2023 Prompt engineering as an important emerging skill for medical professionals: tutorial *J. Med. Internet Res* **25** e50638

Meta AI 2024 https://llama.com/

Mello M M and Guha N 2023 ChatGPT and physicians' malpractice risk *JAMA Health Forum* **4** e231938

NBME 2024 https://abms.org/newsroom/ais-potential-to-transform-assessments/

Ngo B, Nguyen D and vanSonnenberg E 2022 The cases for and against artificial intelligence in the medical school curriculum *Radiol. Artif. Intell.* **4** e220074

Omiye J A, Lester J C, Spichak S, Rotemberg V and Daneshjou R 2023 Large language models propagate race-based medicine *NPJ Digit. Med* **6** 195

OpenAI 2024 https://openai.com/index/hello-gpt-4o/

Perplexity AI 2024 Perplexity AI Pro Version (January 2025), [Claude 3.5 Sonnet model] http://www.perplexity.ai/

Petal 2024 https://www.petal.org/

Pottle J 2019 Virtual reality and the transformation of medical education *Future Healthc. J.* **6** 181–5

Rabie R M 2023 The role of artificial intelligence and personalized education in medical curriculum: a systematic review of applications and challenges *J. Med. Educ.* **20** 123–35

Safranek C W, Sidamon-Eristoff A E, Gilson A and Chartash D 2023 The role of large language models in medical education: applications and implications *JMIR Med. Educ.* **9** e50945

Sauder M, Tritsch T, Rajput V, Schwartz G and Shoja M M 2024 Exploring generative artificial intelligence-assisted medical education: assessing case-based learning for medical students *Cureus* **16** e51961

Scarfe P, Watcham K, Clarke A and Roesch E 2024 A real-world test of artificial intelligence infiltration of a university examinations system: a 'turing test' case study *PLoS One* **19** e0305354

Singhal K, Azizi S, Tu T *et al* 2023 Large language models encode clinical knowledge *Nature* **620** E19

Singhal K *et al* 2023 Large language models encode clinical knowledge *Nature* **620** 172–80 (Publisher correction)

Sorin V, Brin D, Barash Y *et al* 2024 Large language models and empathy: systematic review *J. Med. Internet. Res.* **26** e52597

Tlili A, Shehata B, Adarkwah M A *et al* 2023 What if the devil is my guardian angel: ChatGPT as a case study of using chatbots in education *Smart Learn. Environ.* **10** 15

Vaswani A, Shazeer N, Parmar N *et al* 2017 Attention is all you need arXiv.org https://arxiv.org/abs/1706.03762

Verge 2024 https://www.theverge.com/2024/10/27/24281170/open-ai-whisper-hospitals-transcription-hallucinations-studies

Wang Y, Wang Z, Liu G *et al* 2022 Application of serious games in health care: scoping review and bibliometric analysis *Front. Public Health* **10** 896974

Wang L K, Paidisetty P S and Cano A M 2023 The next paradigm shift? ChatGPT, artificial intelligence, and medical education *Med. Teach* **45** 925

Wang L, Chen X, Deng X *et al* 2024 Prompt engineering in consistency and reliability with the evidence-based guideline for LLMs *NPJ Digit. Med.* **7** 41

xAI 2024 Bringing grok to everyone https://x.ai/news/grok-1212

Yau Y W, Li Z, Chua M T, Kuan W S and Chan G W H 2021 Virtual reality mobile application to improve videoscopic airway training: a randomised trial *Ann. Acad. Med. Singap* **50** 141–8

Yang Y, Liu X, Jin Q, Huang F and Lu Z 2024 Unmasking and quantifying racial bias of large language models in medical report generation *Commun. Med. (Lond.)* **4** 176

Yogarajan V, Dobbie G and Keegan T T 2024 Debiasing large language models: research opportunities *J. R. Soc. N. Z.* **55** 372–95

Zirpe K G, Seta B, Gholap S *et al* 2020 Incidence of medication error in critical care unit of a tertiary care hospital: where do we stand? *Indian J. Crit. Care Med.* **24** 799–803

IOP Publishing

Predictive Analytics in Healthcare, Volume 2
Transforming the future of medicine
Vinithasree Subbhuraam

Chapter 3

Digital transformation in complementary and alternative medicine

Complementary and alternative medicine (CAM) refers to medical products and practices not part of standard medical care. As the name indicates, complementary medicine is used alongside standard medical treatment, while alternative medicine is used instead of standard treatment. The most common reasons for difficulty integrating CAM into traditional practices are the lack of efficacy evidence generated via high-quality clinical trials, standardized protocols around treatment, and regulatory clearance ensuring the safety and efficacy of the procedures. This chapter answers the question—How can artificial intelligence (AI) modernize CAM approaches and improve adoption? Besides listing and elaborating on ten ways AI can improve adoption, this chapter also presents some key case studies of real-world applications of AI in CAM, highlights some companies in this space, discusses the challenges and ethical issues in using AI in CAM and finally concludes by providing some key calls to action for stakeholders.

3.1 Introduction

3.1.1 Overview of complementary and alternate medicine

CAM refers to medical products and practices not part of standard medical care. As the name indicates, complementary medicine is used alongside standard medical treatment, while alternative medicine is used instead of standard treatment. CAM is generally comprised of the following five categories: (1) whole medical systems, (2) mind–body medicine, (3) biologically-based practices, (4) manipulative and body-based practices, and (5) energy medicine.

3.1.1.1 Whole medical systems
Ayurveda: Ayurveda is a traditional medical system that originated in India over 4000 years ago. The core idea of Ayurveda is that disease manifests when there is

doi:10.1088/978-0-7503-2317-8ch3
3-1

an imbalance in the body's life force, called *prana*. The prescriptions by an Ayurvedic practitioner are aimed at restoring this balance—bringing about an equilibrium of the three bodily qualities/doshas called the vata, pitta, and kapha. These treatments generally use dietary changes, herbs, massage, movement, meditation, and other detoxification practices using enemas, oil massages, or nasal lavage.

Homeopathy: Homeopathy was developed in Germany in the late 1700s and is based on the principle that 'like cures like.' That is, any substance that causes disease when consumed in large quantities has the potential to cure the same disease when given in a minute dose.

Traditional Chinese medicine (TCM): TCM is a traditional medical system that originated in China. It is based on the belief that the body's vital energy (qi) flows along channels (meridians) to keep our health in balance. When disease occurs, TCM aims to restore the balance between two forces—yin and yang—which could manifest in the body as cold and heat, deficiency and excess, and internal and external.

Naturopathic medicine: It is a system that uses natural elements such as air, water, light, heat, and massage, along with herbal products, nutritional guidance, and aromatherapy to stimulate the body to heal itself.

3.1.1.2 Mind–body medicine
Mind–body medicine is based on the belief that regulating mental and emotional factors is essential to improving physical health. It incorporates behavioral, psychological, and spiritual techniques such as breathing (meditation), mental focus (biofeedback, hypnosis, imagery, art, music, dance, etc), and body movements (yoga, tai chi, etc) to help relax the body and mind and stimulate healing.

3.1.1.3 Biologically based practices
This type of CAM utilizes natural biologics such as vitamins, supplements, botanicals, and other special diet therapies.

3.1.1.4 Manipulative and body-based practices
This practice offers techniques focused primarily on the body's structures, such as bones and joints. Examples include massage therapy, chiropractic therapy, cupping, acupuncture, and reflexology.

3.1.1.5 Energy healing
This practice is based on the concept that universal life force or subtle energy fields called biofields exist in and around the body and affect health. By manipulating these fields via techniques such as Reiki and therapeutic touch, healing is stimulated.

3.1.2 CAM's role in holistic health and wellness

Western medicine generally focuses on disease-specific treatment, often targeting specific organs or systems using cutting-edge technologies and evidence-based practices. Treatments are faster compared to CAM but may have side effects. Western medicine is phenomenal for treating acute, life-threatening conditions such as infections, trauma, organ failure, etc. On the other hand, CAM practices take a holistic approach addressing the mind, body, and spirit. They focus on the root causes of the imbalances rather than just the symptoms. They generally follow the principle: 'A human being is more than the sum of their parts. Treat the whole person, not just the illness.' Practitioners often take time to understand the patient's lifestyle and emotional and spiritual needs and provide personalized care. They encourage the active participation of the patients in their care rather than just recommending pills and unnecessary surgeries. Many recommendations for managing and treating chronic conditions include lifestyle changes, stress reduction tools, and dietary changes to strengthen the body's natural immune response. Since the root cause is addressed via healthy habit formations, the probability of the recurrence of disease is reduced, and so is the financial burden associated with sickness.

It is evident that both Western and alternative medicine address the health and well-being of the patient, but in different ways via differing philosophies, methodologies, and applications. Over the years, it has been encouraging to see that many healthcare providers follow an integrative approach, combining the strengths of both systems. An example would be a provider/center that offers cancer treatment using chemotherapy alongside yoga, meditation, and acupuncture to manage side effects and improve patient outcomes.

3.1.3 Adoption issues

Though CAM practices could play a significant role in health and wellness, several limitations and challenges must be addressed for successful integration with the traditional healthcare system. The most common reasons for difficulty integrating CAM into traditional practices are the lack of efficacy evidence generated via high-quality clinical trials, the lack of standardized protocols around treatment, and the lack of regulatory clearance ensuring the safety and efficacy of the procedures. Because of these lacks, some formulations used in these practices have been found to contain heavy metals (Saper *et al* 2004, Mukhopadhyay *et al* 2021) and to cause heavy metal toxicity (Mikulski *et al* 2017), leading to skepticism among mainstream medical professionals and patients. Besides, CAM practices are deeply rooted in cultural traditions, which complicate the creation of universal guidelines for CAM integration into global healthcare systems. Most patients are unaware of CAM options or may hold misconceptions about their benefits and risks. Even if physicians or patients want to try some practices, insurance does not cover most therapies.

3.1.4 Can AI improve CAM adoption?

In the current changing landscape of healthcare due to the dominance of technology, CAM practices have much catching up to do. AI has the potential to revolutionize CAM adoption by playing a significant role in addressing each of the limitations and challenges described in the previous section (Ng *et al* 2024). The rest of this chapter focuses on how AI can play this role (section 3.2), presents some key case studies of real-world applications of AI in CAM (section 3.3), and highlights some companies in this space (section 3.4). Section 3.5 discusses the challenges and ethical issues in using AI in CAM. Section 3.6 concludes this chapter by providing some key calls to action for stakeholders.

3.2 The intersection of AI and CAM

This section presents ten ways AI can address challenges associated with CAM's credibility, accessibility, personalization, and scalability (figure 3.1).

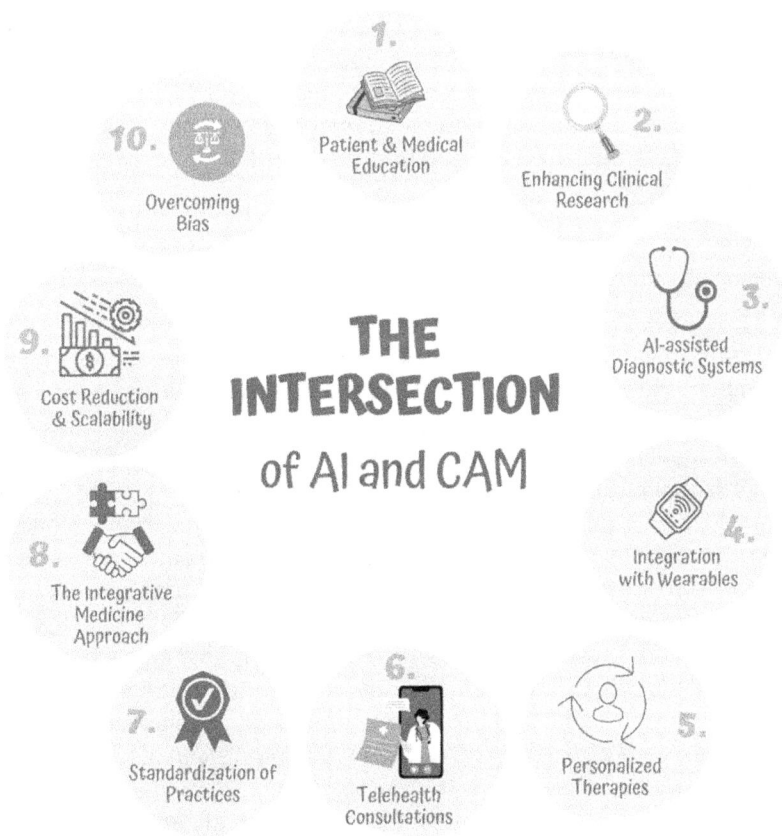

Figure 3.1. Ten ways AI can address challenges associated with CAM's credibility, accessibility, personalization, and scalability.

3.2.1 Patient and medical education

Education is the first step in improving awareness and understanding of alternative medicine. With the advent of large language models (LLMs), it is now easier to create accessible educational materials, help patients interpret complex medical information, and enhance communication between patients and healthcare providers, making them accessible to broader audiences (Aydin *et al* 2024). This is not only for patients, LLMs can also prove to be very useful in CAM education training. They can generate personalized learning materials, create interactive simulations, provide adaptive assessments, answer complex medical questions, and even generate realistic patient scenarios for practice (Abd-Alrazaq *et al* 2023). This allows for a more engaging and tailored learning experience for aspiring CAM students and healthcare professionals.

AI chatbots can also be beneficial in answering questions about CAM practices and improving physician knowledge, patient understanding, and trust. Kim *et al* (2023) recently published a commentary in the Integrative Medicine Research Journal on how ChatGPT could be used in the CAM field. They tested ChatGPT-4 regarding commonly used CAM interventions for cancer-related fatigue. They observed that its response about evidence-based recommendations for CAM interventions was acceptable and in line with the latest findings. However, when more detailed information on the treatment regimen was requested, they found some missing details that make us question if ChatGPT can deliver actionable and reliable knowledge yet.

3.2.2 Enhancing clinical research

One key reason CAM practices struggle with mainstream acceptance is the lack of robust scientific evidence. To address this issue, AI can be extremely useful in streamlining clinical research, optimizing clinical trials, and assisting in conducting high-quality, large-scale trials. Most current studies have shown that AI could be a potential tool to identify and recruit suitable trial participants by studying electronic health records and ensuring the selection of diverse and representative populations (Askin *et al* 2023). Other opportunities for AI in clinical trial design include reducing sample sizes (Krittanawong *et al* 2019). By analyzing electronic medical records, AI can predict which patients are likely to drop from a clinical trial. Efforts can then be made to provide additional education and follow-up to these patients to ensure longer participation, thereby retaining enrolled participants and reducing the need for a larger sample size. AI can help in clinical trial design—more accurate hypothesis generation, cohort composition, monitoring, adherence, and endpoint selection (Harrer *et al* 2019, Delso *et al* 2021). Such tools can also help in designing adaptive trials where adjustments to trial parameters are made using real-time analysis during trials. AI can also potentially create virtual control groups (synthetic control arms) to reduce the need for placebos in trials where withholding treatment might be ethically difficult (Lee and Lee 2020).

3.2.3 AI-assisted diagnostic systems

Data is the fuel for AI. If structured data collection systems are in place to collect clinical, anecdotal, outcome-related, and real-world evidence data, AI models can easily be created to analyze these vast amounts of data to validate the efficacy of CAM therapies and also to develop AI-assisted clinical decision support and diagnostic systems. Machine learning, deep learning, and generative AI models can learn the relationships in complex, multivariate datasets, images, and other data pools from traditional and CAM medicine patient histories. Chu *et al* (2022) reviewed the use of AI in CAM. They reviewed 32 studies that fell into three major categories (acupuncture, tongue and lip diagnoses, and herbal medicine). They concluded that there is potential in using AI-based diagnosis and screening in CAM. However, more large-scale clinical trials are necessary to gather sufficient evidence on their successful use and acceptance by the CAM community. Feng *et al* (2021) highlight the benefits of using AI technology in various TCM modalities such as acupuncture, Tui Na massage, and Qigong. Duan *et al* (2021) have done a similar study to analyze the use of AI in TCM in China and abroad, highlighting use cases in telemedicine, TCM education, and intelligent syndrome differentiation. Zhang *et al* (2020) developed an AI-based assistive diagnostic system to diagnose multiple diseases in TCM (187 disease types) by analyzing 22 984 electronic health record notes. The disease-type prediction accuracies of the top one, top three, and top five were 80.5%, 91.6%, and 94.2%, respectively. Tian *et al* (2023) showed that AI-based research on the four TCM diagnostic methods (inspection, listening, smelling, inquiring, and palpation) is rapidly developing.

3.2.4 Integration with wearable devices

Over the past decade, wearable technologies have transformed healthcare by enabling continuous, real-time monitoring of vital signs, physical activity, and other health metrics. These devices were initially developed and used for fitness tracking. They have evolved to include advanced medical applications, such as detecting arrhythmias with ECG-enabled smartwatches (Pay *et al* 2023), monitoring glucose levels in diabetic patients through non-invasive sensors (Tang *et al* 2020), managing chronic diseases with wearable devices (Jafleh *et al* 2024), etc. Integration with AI has amplified their impact as now the data can be analyzed to obtain personalized health insights, detect conditions early, and deliver telehealth services. A person following CAM therapies and using wearable devices could contribute objective data that could be analyzed to study therapy outcomes. For example, a wearable device could measure stress reduction after a mindfulness session or acupuncture treatment, providing quantifiable evidence.

3.2.5 Prediction and personalization of CAM therapies

CAM practices often lack a one-size-fits-all approach, making personalization crucial for effectiveness. With personalized patient data, such as individual health and lifestyle profiles, genetic data, microbiome data, medical history data, Western medicine data, and lifestyle data, personalized disease management and treatment plans can be established using AI. For example, AI can help develop effective personalized nutrition plans based on an individual's gut microbiome data. Another example could be that chatbots and virtual assistants can guide individuals through personalized meditation or yoga routines based on real-time stress levels detected through wearables. One of the hallmarks of alternative medicine is its over-emphasis on prevention rather than cure. Predictive models powered by AI and personalized datasets from individual patients can identify future disease risks and suggest preventative protocols to reduce or eliminate risk. For example, such models could suggest stress management practices such as yoga for stress-prone individuals and herbal remedies for individuals with genetic predispositions to inflammation.

3.2.6 Telehealth services for CAM consultations

Preventive health and wellness and routine care are key to reducing disease burden. CAM, being considered and used more for preventive health, could be valuable if there is access equity, especially in rural or underserved areas. The COVID-19 pandemic accelerated the adoption of CAM for remote consultations. Der-Martirosian *et al* (2023) examined the implementation of complementary and integrative health-based therapies in 18 Veterans Health Administration sites between 2019 and 2021. They found that televisits comprised 53.7% of self-care therapies in 2020 and increased to 82.1% in 2021. Implementation challenges such as technical difficulties, limited technology access, capacity restrictions, and virtual visit tracking issues were noted. AI could effectively handle some operational issues related to visit and capacity tracking and optimization. Shah *et al* (2023) analyzed 62 articles on CAM delivered through telemedicine. They found that the greater accessibility of CAM therapies via telemedicine could improve clinical outcomes, decrease healthcare spending, and enhance patient satisfaction regarding managing mental health-related and chronic diseases. Besides addressing operational challenges in telehealth implementation, AI tools such as chatbots could also help in symptom analysis and initial consultations. AI-powered virtual assistants can provide personalized guidance to users through yoga, breathing, or mindfulness exercises (Niles *et al* 2012).

3.2.7 Standardization of CAM practices

CAM practices often lack standardization, leading to inconsistent practices and skepticism. Unrelated therapies can originate from different geographic regions; to date, these therapies have not been standardized across global regions. A basic infrastructure should be set up for large-scale global data and methodology collection, and AI can be utilized to address variability in diagnoses, treatments, and outcomes. AI can also create a unified database of herbal medicines and pharmaceutical drugs and study dosages and interactions, thereby ensuring patient safety. On the education front, AI can play an important role in developing standardized training modules and certification programs. It can also be used to develop standardized disease screening and diagnostics protocols that integrate multiple healthcare systems, including Western medicine.

3.2.8 The integrative medicine approach

For the most part, CAM and traditional medicine often function in silos, limiting the benefits of an integrative approach. Bridging the gap between the two requires a multifaceted approach; the key is developing evidence-based frameworks that combine both approaches. CAM approaches should have evidence-based validations via thorough and rigorous scientific evaluations and implementation of quality control mechanisms. AI-powered studies connecting the impact of integrative approaches should also be conducted, and unified documentation systems should be created. Then, certifications and regulatory approvals should be obtained, and standardized practice protocols should be created. Once these aspects are in place, AI tools can assist in developing learning systems for integrative approaches to be taught to medical students and physicians in traditional medical schools to help them improve their skills.

On the patient front, physicians should be trained to consider the whole person, not just the symptoms, and present integrative treatment plans to their patients, empowering patients in their healthcare decisions. Practitioners should also maintain cultural sensitivity and respect for all approaches and not push one over the other. On the technology front, digital health platforms that combine both approaches should be developed. Such platforms could also include data from wearable devices. Electronic health records and telemedicine platforms should be updated to capture and utilize traditional and alternative medicine-based treatments. One powerful example of an integrative approach is to recommend acupuncture for pain management along with conventional physical therapy. Even though some healthcare institutions offer such treatment plans now, they are not well documented and followed up to understand the impact. A recent report by Kearney highlights all these aspects (Kearney Report 2024).

3.2.9 Cost reduction and scalability

Many CAM therapies are not covered by insurance, making accessibility and cost-related nonadherence an issue for the population that needs them the most (Zhang and Meltzer 2021). There are two ways to address this issue. One is to lower the operational cost for providers, such as using AI to automate routine tasks such as initial health assessments, to reduce the total cost of consultations. The other is to build enough evidence of the preventative nature of these therapies so that value-based insurance companies can cover these therapies. Scalable AI-powered digital health platforms that can be downloaded for a fee or subscription can also help large populations access these alternative therapies at an affordable price. Examples include AI apps offering guided personalized yoga or diet plans. The financial burden can also be reduced by integrating evidence-supported complementary therapies into traditional care (Bhoo-Pathy *et al* 2021, Kong *et al* 2022a).

3.2.10 Overcoming bias

There are several perceived myths and misconceptions surrounding CAM that can deter some patients. As described throughout this section, it is evident that if we integrate AI-driven approaches into complementary, alternative, and integrative medicine to generate evidence and validation, it is possible to make these practices more credible, effective, and widely available. Alternative medicine, which generally refers to practices such as herbal supplements and detox diets meant to replace conventional treatments, is generally unproven and not supported by scientific evidence. Their interactions with conventional treatments are also not established. So, it makes sense to be wary of such treatments.

However, integrative medicine, which uses methodologies such as yoga, massage, music therapy, etc, to complement conventional medicine, has garnered much scientific evidence over the years. For example, a study conducted at the MD Anderson Cancer Center in 2019 (Garcia *et al* 2019) found that patients who received acupuncture during radiation treatments for head and neck cancers experienced significantly less dry mouth. In another study (Chaoul *et al* 2018), it was observed that yoga helped reduce insomnia in women when they received chemotherapy for breast cancer. There are many such evidence-based supportive studies for the integrative medicine approach of treating patients by considering the whole person —their physical, mental, emotional, and social health. AI can fast-track evidence generation to support integrative medicine and, eventually, some elements of alternative medicine. AI-powered tools utilizing natural language processing (NLP) can study patient reviews and testimonials to capture evidence of positive outcomes and address myths or misconceptions by providing accurate information.

3.3 Case studies of applications of AI in CAM

3.3.1 AI in TCM pulse diagnosis

TCM doctors use four diagnostic methods to understand the patient's constitution and syndromes. They are inspection, auscultation and olfaction, inquiry, and palpation. Leung *et al* (2021) reviewed the use of AI in TCM sphygmopalpation or arterial pulse diagnosis. Digitizing this procedure is not easy because the recognition of arterial pulse characteristics is highly dependent on the experiences and feelings of TCM practitioners. To accurately mimic these practitioners' experiences, the device should be able to apply different pressure levels, measure the pulses at specific points, and learn to classify 28 types of arterial pulses. AI could be more valuable in this procedure since it involves big data and complex variables. They describe the results of several studies that utilize machine learning or deep learning in TCM diagnosis and treatment (Zhao and Small 2005, Xu *et al* 2007, Chen *et al* 2009, Luo *et al* 2018, Liang *et al* 2019, Zhang *et al* 2019, Chen *et al* 2020).

Even though AI seems to be a powerful tool showing good accuracy in AI-assistive TCM pulse diagnosis studies, the study lists several challenges:

- The correlation between the subjective finger sensations of an experienced TCM practitioner and objective machine-recorded pulse wave variables might not be consistent. Different practitioners might interpret the same pulse characteristics differently, and even the same practitioner might interpret the same pulse waves in the same patient differently based on circumstances. A lot of data is required from multiple patients and practitioners under various circumstances to train AI models more accurately.

- The algorithms will not be accurate if there are differences in the time, pressures, and positions of pulse data.
- Pressure-sensing wearables not customized for TCM pulse diagnosis might not provide accurate data acceptable to TCM practitioners.
- The doctors wanted real-time measurement of the machine-recorded pulse waves.

The authors also offer two solutions to address the above issues:
- Development of TCM pulse diagnostic systems based on TCM principles: Individual diagnostic systems could be first developed by training the algorithms using data from patients and subjects from individual physicians. Then, AI systems can compare and analyze the consistency of diagnosis among different physicians (having the same expertise and work experience). Physicians can gather to re-analyze any inconsistent interpretations to resolve potential differences. Their findings are fed to the AI system for them to understand the consistent and inconsistent components of pulse recognition by different doctors. This is one way of standardizing and digitizing the TCM pulse diagnosis process.
- Development of real-time AI data collection and analysis systems: The proposed solution was to collect data remotely and transport the pulse data acquired at the same positions as the TCM practitioners.

As a result, their research group recently reported the development of a pulse-sensing platform for studying and analyzing arterial pulse patterns using TCM (Kong *et al* 2022b). The platform consisted of a palpation robotic hand comprising three robotic fingers for pulse measurement and a neural network algorithm for pulse pattern recognition. On testing, the platform obtained 97.4% accuracy in identifying three consistent pulse patterns described by TCM doctors.

3.3.2 Modernizing Ayurveda using AI

Several studies have explored AI integration in multiple Ayurveda branches (Nesari 2023, Buvana *et al* 2024, Ranade 2024). They are described below:
- **Diagnostics (*Roga Nidana*)**: AI can learn from years of medical history and investigations to uncover new patterns that human practitioners might have missed.
- **Drug discovery and pharmacological research (*Dravya Guna*)**: Generative tools like LLMs can be effectively trained on various ancient literature, research papers, and clinical data to identify potential medical herbs and their benefits. For example, Mohanraj *et al* (2018) presented IMPPAT (Indian medical

plants, phytochemistry and therapeutics), a manually curated database of 1742 Indian medicinal plants, 9596 phytochemicals, and 1124 therapeutic uses spanning 27 074 plant-phytochemical associations and 11 514 plant-therapeutic associations. They used data mining techniques to filter a subset of 960 potential druggable phytochemicals, most of which do not have similarities with existing FDA-approved drugs. These phytochemicals could be developed into prospective drugs. AI can also help optimize drug development by predicting herb–drug interactions and minimizing adverse side effects.

- **Quality control (in *Rasashastra and Bhaishajya*)**: *Rasashastra* is the study of inorganic pharmaceutical preparations. *Bhaishajya Kalpana* is the pharmaceutical department that deals with the preparation of herbal and organic pharmaceutical preparations. Often, there is a lack of standard protocols and quality control in the preparation of Ayurvedic formulations, which leads to safety and efficacy concerns. AI can help monitor manufacturing processes and detect any deviations from standards.
- **Prakriti assessment (*Swasthavritta*)**: *Swasthavritta* is a branch of Ayurveda that deals with disease prevention through a holistic approach using yoga, diet, and hygiene. In Ayurveda, treatments are personalized based on the patient's Prakriti, which is determined by the presence of three Doshas. Simple AI algorithms such as Naïve Bayes have been used to build a Prakriti assessment chatbot (Kulkarni *et al* 2024).
- **Personalized treatment plans (*Kaya Chikitsa*)**: Combined with these Prakriti assessments and data from wearables, AI can be powerful in presenting personalized interventions.
- **Support surgery (*Shalya Tantra*)**: AI-powered surgical robots can assist in precise surgical procedures, such as Agni Karma (thermal cauterization treatment that uses heat to treat pain) to reduce human error. Such robots can also help train medical students.
- **Support women's health (*Stri Roga (gynecology) and Prasuti (obstetrics)*)**: AI-powered applications can track menstrual cycles and hormonal patterns to diagnose and treat reproductive issues in women.
- **Support children's health (*Kaumarabhritya*)**: Predictive analytics can analyze data related to a child's growth patterns and development milestones to identify issues early.
- **Medical education (*Samhitha Siddhanta*)**: Samhitha Siddhanta is a department of Ayurveda that studies the basic principles. AI can analyze classic Ayurveda texts, correlate the knowledge captured with modern medical research, and create learning systems, case studies, and assessment tools to aid medical education.
- **Patient education:** AI can easily create customized educational content and support patient engagement in procedures such as Panchakarma.

AI can have a powerful impact on modernizing the traditional wisdom of Ayurveda for improved health outcomes. Some companies are already developing AI-based tools for Ayurvedic practice:

Nadi Tarangini (Nadi Tarangini 2025) has developed an AI-based tool for Nadi Pariksha that captures pulse data (using ultra-sensitive sensors), analyzes it, and generates a report of 22 Ayurvedic parameters. The system can also recommend a personalized diet and yoga regimen, guiding users on their *Ritucharya* (seasonal regimen) and *Dinacharya* (daily routine) practices.

Ayur AI (Ayur AI 2025) is a deep tech company that uses advanced digital and blood biomarkers, genomics, and AI algorithms to provide personalized Ayurveda wellness and disease management plans.

Kama Ayurveda (Kama Ayurveda 2025), an Ayurvedic beauty company, has partnered with Holition Beauty, an AI and augmented reality (AR) provider, to utilize facial scans for Ayurvedic dosha consultation done from the comfort of the patient's home.

3.4 Some notable companies in integrative medicine

A few companies are at the forefront of developing digital-driven solutions for integrating traditional and alternative medicine. Some of them are listed below.

Quantum Meta Health (Quantum Meta Health 2025) uses cutting-edge quantum physics principles to offer remote body scans and holographic treatments using 8D nonlinear systems life resonance intelligence systems. It uses DNA and RNA to find conditions by understanding the wave characteristics of various body parts and bio-resonance technology to remove energy blocks that cause disease.

Core Spirit (Core Spirit 2025) is a social platform focusing on human and planetary enhancement through diverse online and offline media that unite content, experts, and practitioners. Users can, therefore, choose from a blend of disease management and treatment options.

Sofia Health (Sofia Health 2025), similar to Core Spirit, is another platform that brings together traditional, integrative, nutritional, holistic, beauty, fitness, and spiritual healing professionals.

Being Health (Being Health 2025) is an integrative mental health practice that takes a whole-person approach to mental health. Their patients have access to psychiatrists, psychotherapists, functional medicine providers, nutritionists, and acupuncturists, as well as novel treatments like ketamine therapy.

Calcium Health (Calcium Health 2025) has developed a digital health Software as a Service platform for integrated medicine providers and patients that utilizes tools and analytics to improve patient engagement, communication, remote patient monitoring, and care team productivity.

Readers can visit the FS6 website to learn more about other companies changing the alternative medicine landscape (FS6 2025).

3.5 Challenges

Technology integration into traditional practices always comes with a variety of challenges. With respect to AI use, significant issues remain to be addressed. Some of them are described below:

- **Data collection and standardization**

  One of the greatest challenges in incorporating AI within the CAM field is the lack of structured, complete datasets. Another issue is the lack of standardization of terms used in several CAM modalities, making it difficult to develop standardized electronic medical records.

- **Bias in AI models**

  AI models are known to rely completely on the quality of data input during training. The resultant predictive model or LLM will be biased if there is no cultural and regional diversity in training datasets. Another layer of bias that needs to be avoided is the bias toward Western medicine while integrating CAM practices.

- **Inherent issues of LLMs**

  One of the key issues of LLMs is hallucinations, where they generate inaccurate, irrelevant, and sometimes nonsensical responses. Another issue is the lack of consistency in responses—the same prompts generate different results. Thirdly, a LLM's knowledge base is not current. A detailed overview of the inherent issues of using LLMs in healthcare, specifically in medical education, is provided in chapter 2 of this book. Because of these limitations, the current recommendation is to use these generative AI tools as an augmentative rather than a stand-alone tool. Furthermore, it is critical that expert reviews and cross-referencing of the responses generated by multiple stakeholders are conducted to ensure that accurate and reliable information is garnered.

- **Regulatory and ethical concerns**

  CAM modalities already incorporate diverse therapies and practices, but many are still not regulated. Adding AI to this already challenging aspect introduces more complexities in regulating these products. Clear guidelines for obtaining regulatory clearance have to be established. Legal frameworks should be created to address liability issues related to AI-prone errors and adverse outcomes.

- **Data privacy and security**

  CAM practices generally collect more patient-sensitive data, such as lifestyle and emotional health information. Strict data privacy and security guardrails must be in place to gain patient trust in using AI-powered solutions in CAM practices. Patients should also be allowed to learn about these tools and provide informed consent.

- **Language disparities**

  According to OpenAI (OpenAI 2023), the multitask language understanding (MMLU) benchmark test results showed that GPT-4 had an accuracy of 84.1% for English, 77% for Korean, and 62% for Telugu. Ando *et al* 2024 also found this to be true in their experiments where ChatGPT generated different answers depending on the language of the question. This suggests that there is room for improvement in this area. Work is being conducted to address this issue. Tan *et al* (2024) realized that most LLM models are trained on datasets in English and, therefore, struggle to provide accurate answers in Chinese, especially in TCM question-answering (QA) systems. To address this issue, they pre-trained a dialogue model called MedChatZH using curated Chinese medical books and fine-tuned the model with a medical instruction dataset. They observed that this model could outperform several Chinese dialogue baselines.

3.6 Conclusions and future directions

Complementary, alternative, and integrative medicine are slowly and steadily achieving international recognition. The first-ever World Health Organization (WHO) Traditional Medicine Global Summit 2023 conference was held in India from August 17 to 18, 2023 (WHO 2023), to mobilize political commitment and evidence-based action supporting these approaches to medicine. In a commentary on the summit, Liu and Gong (2024) reported that AI could help modernize TCM research methods. This chapter described how AI could do that for TCM and other CAM modalities. The case studies showed immense potential to transform practices in this field of medicine, potentially improving its widespread adoption. However, as highlighted, challenges remain, and more work is needed in this area.

AI-driven CAM can become a cornerstone of a holistic, patient-centric, and sustainable healthcare system, provided we advance the space with a strong collaborative effort among CAM practitioners, technologists, researchers, and policymakers. Here are a few critical areas that require immediate attention:

- Increasing research funding to support large-scale studies to establish the safety and efficacy of CAM practices.
- Building global research hubs to establish standardized data collection platforms for each type of CAM modalities and enable cross-disciplinary research and collaboration. Placing the patient and disease at the center, how can all these modalities come together to develop preventative, diagnostic, management, and treatment protocols?
- Creating interoperable systems to bridge the gap between CAM and conventional Western medicine. For example, AI and digital health tools can standardize CAM data and integrate it with electronic medical records.
- Developing AI-driven wearable devices specific to the practice requirements of each CAM modality.
- Using such wearables and other metrics to build digital twins.
- Creating evidence-based regulatory frameworks to increase practitioner confidence and competence.
- Developing digital platforms to enhance patient and medical student education.

References

Abd-Alrazaq A, AlSaad R, Alhuwail D *et al* 2023 Large language models in medical education: opportunities, challenges, and future directions *JMIR Med. Educ* **9** e48291

Ando K, Sato M, Wakatsuki S *et al* 2024 A comparative study of English and Japanese ChatGPT responses to anaesthesia-related medical questions *BJA Open.* **10** 100296

Ayur AI 2025 https://ayurai.io/

Askin S, Burkhalter D, Calado G and El Dakrouni S 2023 Artificial intelligence applied to clinical trials: opportunities and challenges *Health Technol. (Berl.)* **13** 203–13

Aydin S, Karabacak M, Vlachos V and Margetis K 2024 Large language models in patient education: a scoping review of applications in medicine *Front. Med. (Lausanne)* **11** 1477898

Being Health 2025 https://beinghealth.co/

Bhoo-Pathy N, Subramaniam S, Khalil S *et al* 2021 Out-of-pocket costs of complementary medicine following cancer and the financial impact in a setting with universal health coverage: findings from a prospective cohort study *JCO Oncol. Pract* **17** e1592–602

Buvana V M, Mutalikdesai V, Sajitha K, Deepthi R and Balakrishnan A 2024 A bird's eye view on the integration of artificial intelligence (AI) in Ayurveda *J. Ayurveda Integr. Med. Sci.* **9** 167–72

Calcium Health 2025 https://calciumhealth.com/integrative-medicine/

Chaoul A, Milbury K, Spelman A *et al* 2018 Randomized trial of Tibetan yoga in patients with breast cancer undergoing chemotherapy *Cancer* **124** 36–45

Chen Y, Zhang L, Zhang D and Zhang D 2009 Wrist pulse signal diagnosis using modified Gaussian models and Fuzzy C-Means classification *Med. Eng. Phys.* **31** 1283–9

Chen J, Huang H, Hao W and Xu J 2020 A machine learning method correlating pulse pressure wave data with pregnancy *Int. J. Numer. Method Biomed. Eng* **36** e3272

Chu H, Moon S, Park J, Bak S, Ko Y and Youn B Y 2022 The use of artificial intelligence in complementary and alternative medicine: a systematic scoping review *Front. Pharmacol* **13** 826044

Core Spirit 2025 https://corespirit.com/

Der-Martirosian C, Shin M, Upham M L, Douglas J H, Zeliadt S B and Taylor S L 2023 Telehealth complementary and integrative health therapies during COVID-19 at the U.S. department of veterans affairs *Telemed. J. E-Health* **29** 576–83

Delso G, Cirillo D, Kaggie J D, Valencia A, Metser U and Veit-Haibach P 2021 How to design AI-driven clinical trials in nuclear medicine *Semin. Nucl. Med.* **51** 112–9

Duan Y Y, Liu P R, Huo T T, Liu S X, Ye S and Ye Z W 2021 Application and development of intelligent medicine in traditional Chinese medicine *Curr. Med. Sci.* **41** 1116–22

Feng C, Zhou S, Qu Y *et al* 2021 Overview of artificial intelligence applications in Chinese medicine therapy *Evid. Based Complement. Altern. Med* **2021** 6678958

FS6 2025 https://f6s.com/companies/alternative-medicine/united-states/co

Garcia M K, Meng Z, Rosenthal D I *et al* 2019 Effect of true and sham acupuncture on radiation-induced xerostomia among patients with head and neck cancer: a randomized clinical trial *JAMA Netw. Open* **2** e1916910

Harrer S, Shah P, Antony B and Hu J 2019 Artificial intelligence for clinical trial design *Trends Pharmacol. Sci.* **40** 577–91

Jafleh E A, Alnaqbi F A, Almaeeni H A, Faqeeh S, Alzaabi M A and Al Zaman K 2024 The role of wearable devices in chronic disease monitoring and patient care: a comprehensive review *Cureus* **16** e68921

Kama Ayurveda 2025 https://holition.com/work/kama-ayurveda

Kearney Report 2024 https://middle-east.kearney.com/industry/health/article/integrating-alternative-medicine-into-conventional-health-systems

Kim T H, Kang J W and Lee M S 2023 AI Chat bot—ChatGPT-4: a new opportunity and challenges in complementary and alternative medicine (CAM) *Integr. Med. Res* **12** 100977

Kong Y C, Kimman M, Subramaniam S D, Yip C H, Jan S, Aung S *et al* 2022a Out-of-pocket payments for complementary medicine following cancer and the effect on financial outcomes in middle-income countries in southeast Asia: a prospective cohort study *Lancet Oncol.* **10** e416–28

Kong K W *et al* 2022b Sphygmopalpation using tactile robotic fingers reveals fundamental arterial pulse patterns *IEEE Access* **10** 12252–61

Krittanawong C, Johnson K W and Tang W W 2019 How artificial intelligence could redefine clinical trials in cardiovascular medicine: lessons learned from oncology *Per. Med.* **16** 83–8

Kulkarni K, Kota R, Thorat R, Pamnath S, Amrutam V and Raut S 2024 Survey paper on Chatbot for Prakriti assessment *Int. J. Eng. Technol.* **11** 1–8

Lee C S and Lee A Y 2020 How artificial intelligence can transform randomized controlled trials *Transl. Vis. Sci. Technol* **9** 9

Leung Y L A, Guan B, Chen S, Chan H, Kong K, Li W and Shen J 2021 Artificial intelligence meets traditional Chinese medicine: a bridge to opening the magic box of sphygmopalpation for pulse pattern recognition *Digit. Chin. Med* **4** 1–8

Liang Z, Liu J, Ou A, Zhang H, Li Z and Huang J X 2019 Deep generative learning for automated EHR diagnosis of traditional Chinese medicine *Comput. Methods Programs Biomed.* **174** 17–23

Liu X and Gong T 2024 Artificial intelligence and evidence-based research will promote the development of traditional medicine *Acupunct. Herb. Med* **4** 134–5

Luo Z Y, Cui J, Hu X J, Tu L P, Liu H D, Jiao W *et al* 2018 A study of machine-learning classifiers for hypertension based on radial pulse wave *BioMed. Res. Int.* **2018** 2964816

Mikulski M A, Wichman M D, Simmons D L, Pham A N, Clottey V and Fuortes L J 2017 Toxic metals in ayurvedic preparations from a public health lead poisoning cluster investigation *Int. J. Occup. Environ. Health* **23** 187–92

Mohanraj K, Karthikeyan B S, Vivek-Ananth R P *et al* 2018 IMPPAT: a curated database of indian medicinal plants, phytochemistry and therapeutics *Sci. Rep.* **8** 4329

Mukhopadhyay S *et al* 2021 Heavy metals in Indian traditional systems of medicine: a systematic scoping review and recommendations for integrative medicine practice *J. Altern. Complement. Med.* **27** 915–29

Nadi Tarangini 2025 https://naditarangini.com/

Nesari T M 2023 Artificial intelligence in the sector of Ayurveda: scope and opportunities *Int. J. Ayurveda Res* **4** 57–60

Ng J Y, Cramer H, Lee M S and Moher D 2024 Traditional, complementary, and integrative medicine and artificial intelligence: novel opportunities in healthcare *Integr. Med. Res* **13** 101024

Niles B L, Klunk-Gillis J, Ryngala D J, Silberbogen A K, Paysnick A and Wolf E J 2012 Comparing mindfulness and psychoeducation treatments for combat-related PTSD using a telehealth approach *Psychol. Trauma* **4** 538–47

OpenAI 2023 GPT-4 Technical Report https://cdn.openai.com/papers/gpt-4.pdf 2023

Pay L, Yumurtaş A Ç, S D I *et al* 2023 Arrhythmias beyond atrial fibrillation detection using smartwatches: a systematic review *Anatol. J. Cardiol* **27** 126–31

Quantum Meta Health 2025 https://quantummetahealth.com/

Ranade M 2024 Artificial intelligence in Ayurveda: current concepts and prospects *J. Indian Syst. Med.* **12** 53–9

Saper R B, Kales S N, Paquin J *et al* 2004 Heavy metal content of ayurvedic herbal medicine products *JAMA* **292** 2868–73

Shah A Q, Noronha N, Chin-See R *et al* 2023 The use and effects of telemedicine on complementary, alternative, and integrative medicine practices: a scoping review *BMC Complement. Med. Ther.* **23** 275

Sofia Health 2025 https://sofiahealth.com/

Tan Y, Zhang Z, Li M *et al* 2024 MedChatZH: a tuning LLM for traditional Chinese medicine consultations *Comput. Biol. Med.* **172** 108290

Tang L, Chang S J, Chen C J and Liu J T 2020 Non-invasive blood glucose monitoring technology:a review *Sensors (Basel)* **20** 6925

Tian Z, Wang D, Sun X *et al* 2023 Current status and trends of artificial intelligence research on the four traditional Chinese medicine diagnostic methods: a scientometric study *Ann. Transl. Med* **11** 145

WHO 2023 The First WHO Traditional Medicine Global Summit 2023. https://who.int/news-room/events/detail/2023/08/17/default-calendar/the-first-who-traditional-medicine-global-summit

Zhang J, Su Q, Loudon W G *et al* 2019 Breathing signature as vitality score index created by exercises of Qigong: implications of artificial intelligence tools used in traditional Chinese medicine *J. Funct. Morphol. Kinesiol* **4** 71

Zhang H, Ni W, Li J and Zhang J 2020 Artificial intelligence–based traditional Chinese medicine assistive diagnostic system: validation study *JMIR Med. Inform* **8** e17608

Zhang J X and Meltzer D O 2021 Association between the modalities of complementary and alternative medicine use and cost-related nonadherence to medical care among older Americans: a cohort study *J. Altern. Complement. Med* **27** 1131–5

Zhao Y and Small M 2005 Equivalence between 'feeling the pulse' on the human wrist and the pulse pressure wave at fingertip *Int. J. Neural Syst.* **15** 277–86

Xu L S, Meng M Q H and Wang K Q 2007 Pulse image recognition using fuzzy neural network *2007 29th Annual Int. Conf. of the IEEE Engineering in Medicine and Biology Society (Lyon, France)* (Piscataway, NJ: IEEE) pp 3148–51

IOP Publishing

Predictive Analytics in Healthcare, Volume 2
Transforming the future of medicine
Vinithasree Subbhuraam

Chapter 4

AI-powered solutions for prostate health management—risk assessment, screening, diagnosis, treatment, and management: a review

The landscape of prostate cancer detection and care has rapidly evolved. It has moved from using conventional imaging and radical surgeries to an era of genomics, precision diagnostics, advanced imaging, and targeted treatments. Lately, the role of artificial intelligence (AI) in this field has moved from building traditional machine learning models to developing large language models (LLMs). This chapter presents the latest advancements in AI-driven applications across the continuum of the journey of a patient with prostate cancer, from risk assessment and stratification to screening, diagnosis, treatment, and management.

4.1 Introduction

The prostate is a gland found beneath the bladder in men. It is about the size of a walnut and produces fluid that protects the urethra and helps sperm survive. According to the World Cancer Research Fund (WCRF 2025), prostate cancer (PCa) is the fourth most common cancer globally and the second most common cancer in men. The top five countries by age-standardized rates (ASR) for prostate cancer incidence are Guadeloupe (France), Lithuania, Martinique (France), Norway, and Sweden. Not considering ASR, the top 10 countries with the highest prostate cancer incidence in 2022 are the US, China, Japan, Brazil, Germany, France, UK, Russia, Italy, and India in that order. The higher incidence rates could be largely due to the widespread screening practices and increased life expectancy in these countries. China, the US, and Brazil had the highest number of deaths from prostate cancer in 2022. In general, men of African–Caribbean or African descent seem to have a higher risk, followed by white men and then Asian men. Wang *et al*

doi:10.1088/978-0-7503-2317-8ch4   4-1   © IOP Publishing Ltd 2025. All rights,

(2022) studied incidence and mortality data of prostate cancer from 174 countries from the GLOBOCAN 2020 database to study the associations of the data with the human development index (HDI). They observed a positive association between prostate cancer incidence and HDI and a negative association between mortality and HDI. A more up-to-date detailed 2022 statistics can be found in Bray *et al* (2024).

It has been established that tumor development is controlled by several biological variables such as tumor biology, microenvironment, environmental exposures, hereditary risk, etc. Based on these conditions, a tumor can either be slowly progressive, rapidly progressive, or even an indolent lesion that grows slowly and is unlikely to cause harm if untreated. Disease-based screening and diagnostic scans done for other purposes have led to cancer overdiagnosis and, therefore, to overtreatment (Esserman *et al* 2013). Adenocarcinoma of the prostate is probably a cancerous condition that has the most significant risk of overdiagnosis and overtreatment (Esserman *et al* 2014, Loeb *et al* 2014, McCaffery *et al* 2019, Dunn *et al* 2022). With prostate-specific antigen (PSA) testing, most commonly used for screening, low-grade tumors are often detected in up to 70% of the diagnoses (Islami *et al* 2021). Over 30% of men over the age of 50 and more than 60% over the age of 80 years have microscopic prostate cancer (Welch and Albertsen 2009), but only about 3% die of prostate cancer (Islami *et al* 2021). Indolent tumors (called Gleason Score 6 or GS6) are not known to cause symptoms or metastases (Labbate *et al* 2022) but are subject to invasive monitoring or treatment. Therefore, Eggener *et al* (2022) suggested that GS6 conditions should be relabeled as precancerous lesions to reduce overdiagnosis and treatment. Epstein and Kibel (2022) countered the suggestion by saying that based on morphological and molecular findings, maintaining GS6 as cancerous is necessary. This led to the publication of a surgeon's viewpoint, where Baboudjian *et al* (2023) reminded that pre-biopsy multiparametric magnetic resonance imaging (mpMRI) is becoming more common to reduce overdiagnosis of GS6 disease, and its use has been endorsed by many international guidelines (Mottet *et al* 2021, Schaeffer *et al* 2021). When mpMRI is negative (with a prostate imaging reporting and data system (PI-RADS $\leqslant$ 2) with a low PSA density < 0.15, biopsies are now mostly avoided. They highlighted that overdiagnosis is now not an issue, but overtreatment is, and recommended that active surveillance (AS) of GS6 prostate cancer should be the way forward. AS is now considered the standard of care for all patients with low-risk prostate cancer. More on this is in section 4.3.1.

These studies indicate that it is critical to understand prostate cancer tumor biology and develop more accurate ways for risk assessment, screening, diagnosis, risk stratification, treatment, and management to avoid overdiagnosis and overtreatment. AI, particularly LLMs, is increasingly transforming prostate cancer care because of the availability of large datasets and computing power. It is now easier than ever to integrate clinical, genetic, demographic, and other risk parameters to develop risk-prediction models for sub-groups of people. LLMs can synthesize medical literature, summarize patient data, and provide evidence-based recommendations for personalized screening. AI-powered algorithms can analyze large volumes of medical images from multiparametric MRI and ultrasound scans to detect and localize cancerous lesions with great precision. These algorithms can also more accurately grade tumors. AI-driven liquid biopsy techniques can analyze circulating tumor DNA and biomarkers to detect cancer non-invasively. Machine learning (ML) and deep learning (DL) algorithms can guide treatment recommendations by providing data-driven, personalized therapeutic strategies. AI can also predict responses to surgery, radiation, hormone therapy, and immunotherapy, optimizing treatment plans for better outcomes. These advancements underscore AI's growing role in enhancing prostate cancer care, leading to more precise and effective patient management.

This chapter presents the more recent advancements in using AI in all areas of prostate cancer management. Section 4.2 describes the various risk assessment and screening tools currently in use and highlights AI-based models. Section 4.3 reviews many studies developing AI-supported models for enhancing the existing approaches to PCa diagnosis and detection. A note on techniques to differentiate benign prostatic hyperplasia (BPH) from cancer is also included in this section. PCa treatment selection, planning, and execution are supported by AI, and these approaches are described in section 4.4. Possible LLM applications are presented in section 4.5. Ten FDA-approved AI-ML solutions for prostate cancer management are listed and described in section 4.6. Concluding remarks are presented in section 4.7.

4.2 Risk assessment and screening

Risk assessment is a comprehensive evaluation of a man's likelihood of developing prostate cancer based on factors such as family history, biomarkers, genetics, and medical history. Risk assessment tools help determine who might need further screening tests, such as PSA testing and digital rectal examination (DRE). However, in the literature, population-based risk assessment tools to evaluate lifestyle risk factors for PCa development independent of the PSA test, imaging tests (ultrasound/ MRI), and/or genetics are rare. Some studies have attempted to develop models using lifestyle and other easily accessible and modifiable risk factors using data from large population-based cohort studies. One study (Kim *et al* 2018) used data from 1 179 172 Korean men to develop a preliminary risk-prediction model using epidemiological lifestyle factors such as age, height, body mass index, fasting glucose levels, family history of cancer, the frequency of meat consumption, alcohol consumption, smoking status, and physical activity. The model presented a c-statistic of 0.887 (95% confidence interval (CI): 0.879–0.895). Yeo *et al* (2021) used age, body mass index, cumulative smoking intensity, Type 2 diabetes mellitus, hypertension, and regular physical exercise as variables in their model. They obtained a c-statistic of 0.826 (95% CI: 0.821–0.832). These two studies are based on data from Korean men. More studies are needed in various populations to develop risk-prediction models using epidemiological lifestyle factors. Ziglioli *et al* (2023) reviewed the literature on modifiable risk factors for prostate cancer (independent of whether risk models were developed or not). They noted that there is strong evidence between being overweight/obese and the risk of developing advanced PCa. They also present evidence of the impact of other modifiable risk factors.

Other risk assessment models are based on genetic data. For example, the polygenic risk score (PRS) predicts susceptibility to prostate cancer by summoning up the effects of many common genetic variants (single nucleotide polymorphisms) linked to disease in genomic studies. Another test is the Select MDx test, which is a urine test to screen for biomarkers of prostate cancer. The ExoDx Prostate Test is another urine-based test that evaluates exosomal RNA biomarkers to predict the likelihood of clinically significant prostate cancer (csPCa).

In individual risk predictions, clinicians mostly use PSA tests and/or DRE. Though PSA screening has been shown to reduce PCa mortality, it is known to lead to unnecessary prostate biopsies and overdiagnosis/overtreatment of indolent PCa (Roobol and Carlsson 2013, David and Leslie 2024). It has been observed that PSA sensitivity ranges from 9% to 33%, depending on age and the PSA cut-off values, and up to 91% of individuals with elevated serum PSA levels do not have prostate cancer (Leal *et al* 2018). Therefore, physicians use several other screening tools to more accurately select patients for prostate biopsy to identify patients at risk for csPCa (Alford *et al* 2017). One such tool is the prostate health index (PHI), a blood test combining total PSA, free PSA, and p2PSA. The 4Kscore test is a non-invasive follow-up test that is more specific than PSA or DRE in determining the probability of aggressive prostate cancer. Benign prostate conditions, such as infection or an enlarged prostate, which influence the PSA test, do not generally influence the result of this test. It is used after an abnormal PSA/DRE test if the physician is considering a biopsy. It is calculated using four different factors—Total PSA, Free PSA, Intact PSA, and Human Kallikrein-2 (hK2, a prostate-specific enzyme that can break bonds in proteins).

Several AI-based risk assessment models have recently been developed to predict PCa. In one study (Nitta *et al* 2019), data on continuous changes in the PSA level over the past two years were accumulated in 512 Japanese men who underwent prostate biopsy after PSA screening. They used the age of patients, PSA level (maximum, minimum, median, mean, and variance level), prostate volumes, white blood cell count in urinalysis, and the result of biopsy as input and target to train three different ML models. They observed that the artificial neural network (ANN) model had a better area under the receiver operating characteristic curve (AUC) than PSA density and PSA velocity alone. In another study including over 4500 patients with tPSA < 10 ng ml^{-1}, a PSA-based dense neural network had an AUC of 0.72, which was improved compared to PSA alone, age, free PSA (fPSA), and the ratio of serum fPSA to tPSA (f/tPSA) alone (Perera *et al* 2021). Chen *et al* (2022) used retrospective data from 551 patients and constructed five PCa prediction models (total PSA-based univariate logistic regression (LR) model and four other multivariate ML models). Variables used include age, BMI, hypertension, diabetes, total PSA (tPSA), fPSA, f/tPSA, prostate volume (PV), PSA density (PSAD), neutrophil-to-lymphocyte ratio (NLR), and pathology reports of prostate biopsy. Their multivariate LR model performed best, with an AUC of 0.918 in the test dataset. Tran *et al* (2024) investigated if the presence of metabolic syndrome and other sociodemographic characteristics could be used to predict PCa using ML models. They observed that metabolic syndrome and its components did not contribute to PCa risk, but a first-degree family history of PCa, age, alcohol consumption, and income were risk factors for PCa.

Several limitations exist for the validity of these AI-based models: (1) Most of these studies are retrospective in nature, may be potentially biased geographically and institutionally, and may not have adequate sample size for generalization. (2) Both training and validation datasets come from the same hospitals. External validation is necessary. (3) Imaging-based risk factors have increasingly become more accepted as powerful risk factors pre-biopsy and are excluded from these models. Unless large-scale, international trials and data analyses are conducted, it is hard to adopt these models in practice due to the lack of evidence of the predictive power of these risk factors.

4.3 Prostate cancer diagnosis and risk stratification

4.3.1 AI in prostate MRI

If the PSA levels are elevated, DRE results indicate abnormalities, and/or the additional risk assessment tests described in the previous section show elevated risk, further diagnostic steps are performed to confirm the presence or absence of cancer. Recently, several large prospective trials concluded that using mpMRI prior to biopsy increases the detection of more aggressive csPCa while decreasing the detection of non-aggressive PCa as compared to transrectal ultrasound (TRUS) guided biopsy (Drost *et al* 2019, van der Leest *et al* 2019). An mpMRI can capture a more detailed image of the prostate gland than a standard MRI. Recently, the European Association of Urology (EAU), the American College of Radiology (ACR), and the American Urological Association (AUA) adjusted their guidelines for prostate cancer diagnosis, advising that a mpMRI is acquired prior to biopsy (Rosenkrantz *et al* 2016, Mottet *et al* 2021, Wei *et al* 2023a, 2023b).

As with any MRI, prostate cancer MRI reading and reporting process is complex, requires experienced expert readers, and is limited by a steep learning curve and inter-reader variability (Smith *et al* 2019). Moreover, there is also a large variability in the scan quality of images obtained from different institutions using different scanners. To standardize the acquisition and interpretation of prostate mpMRI, the PI-RADS was developed and proposed (Turkbey *et al* 2019). It is a standardized scoring system with scores ranging from 1 (very unlikely to be cancer) to 5 (highly suspicious of cancer). To assign a PI-RADS score, radiologists must analyze the size, location, signal intensity, and margin characteristics. AI models can substantially address these issues and improve clinical workflows in several areas of prostate MRI. Most published studies have developed computer-aided diagnosis (CAD) and detection models for image interpretation tasks such as suspicious lesions diagnosis and detection and lesion segmentation (Twilt *et al* 2021, Kim *et al* 2023). AI can also help in image acquisition and registration and provide prognostic information to help physicians decide if the patient will benefit from AS. Below is an overview of the various AI applications in prostate MRI analysis.

Image acquisition: A biparametric MRI (bpMRI) of the prostate uses two imaging sequences, typically T2-weighted and diffusion-weighted imaging (DWI). An mpMRI includes those sequences plus additional ones like dynamic contrast-enhanced (DCE) imaging and provides a more detailed image of the gland to enhance PCa detection accuracy. bpMRI tends to be faster and cheaper as it does not require contract injection, so it is the most commonly acquired one. Though there are traditional methods to accelerate image acquisition (such as SENSE (Pruessmann *et al* 1999) and GRAPPA (Griswold *et al* 2002)), the quality of the resulting images is not good, and they are computationally complex to execute (Kim *et al* 2023). DL models help address these issues in bpMRI (Johnson *et al* 2022, Pal and Rathi 2022). AI has also been helpful in mpMRI acquisition as recent studies propose methods to reduce the contrast dose (Gong *et al* 2018) and DCE-MRI image synthesis (Xie *et al* 2022).

Image quality evaluation: It is important to develop standardized methods to assess the image quality of prostate MRI (Giganti *et al* 2020, 2021). Quality images generated in this manner can be combined to form large-scale public datasets that can be effectively used by researchers worldwide. Studies of using AI in prostate MRI imaging quality assessment are still lacking.

Image post-processing: Movement artifacts are prevalent in almost every medical imaging modality, and re-acquiring these images is time-consuming and costly. AI-based image post-processing algorithms can be used to reduce noise and artifacts in these images. This is another area for growth in prostate MRI analysis.

Diagnosis and detection: Twilt *et al* (2021) present a detailed review of over 59 studies that either proposed models for lesion classification (classify manually annotated regions) or for lesion detection (automatically detect and localize PCa lesions for further review). These models utilize either ML or DL algorithms. Another excellent review in this area is by Li *et al* (2022). ML-based approaches for PCa classification generally involve the manual or semiautomatic annotation of suspicious regions in multiparametric or single-sequence MR by expert readers, extracting features from these and then training ML classifiers using the ground truth (prostate biopsy results). The most commonly studied features are imaging features (size, shape, vascularity, and other texture features), clinical features such as PSA density, etc. The quality of pre-processing steps such as image registration, segmentation, feature extraction, and feature selection significantly affect the accuracy of these models. DL models, however, rely less on human input as they can learn the optimal features directly from the raw pixel data from images. A very common DL framework is based on convolutional neural network (CNN). DL models have shown excellent results in predicting csPCa (Kim *et al* 2023). In lesion detection, no prior lesion annotation is required as the AI algorithm automatically classifies the images on a voxel level. So, PCa detection algorithms could aid in automated prostate MRI assessment. The outcome of these algorithms is a probability map for prostate cancer likelihood and PCa segmentations based on a threshold within the probability map. Integration of such models is also common. For example, a multimodal nomogram was developed by integrating radiomics and deep learning features from biparametric MRI, PI-RADS score, and other clinical variables. This model outperformed traditional radiomics and clinical models (Chen *et al* 2025). The promising nature of ML and DL models in PCa have led to several groups organizing grand challenges involving large-scale datasets—PI-CAI (Saha *et al* 2022, 2024), Prostate 158 (Adams *et al* 2022) and PROSTATEx (Armato *et al* 2018) to enable researchers use these heterogeneous datasets for validating new algorithms.

Segmentation: There are several ways AI can enhance image segmentation accuracy. It can guide the segmentation of the entire prostate gland, which is beneficial to guide biopsies and to measure the gland volume for active surveillance. It can also help delineate individual lesions that can be used for targeted biopsies and surgery planning. Localized lesions can also be input into AI models for diagnosis and risk stratification. The support vector machine is a commonly used ML model (Ozer *et al* 2010, Artan and Yetik 2012). DL techniques, especially the U-Net architecture (Bhandary *et al* 2023, Hong *et al* 2023, Shen *et al* 2025) have shown excellent segmentation performance due to their ability to preserve the original image structure. The dice similarity coefficient (DSC) is a metric that compares how similar two images are. Many recent studies have demonstrated a DSC of 0.9 between the predicted segmentation and ground truth segmentation when U-Net architectures were used (Kim *et al* 2023). Segmenting small lesions is always challenging, but algorithm advancements can help address this issue (Tong *et al* 2020).

Registration: MRI images are generally fused with TRUS images to guide targeted biopsies and combined with computer tomography (CT) images for radiation therapy planning. Cross-modality registration is always a challenge in medical imaging. DL-based methods have shown more promising results than traditional techniques such as B-spine-based methods, finite element modeling, etc (Kim *et al* 2023, Wu *et al* 2023). Pre-operative MRI images can also be registered with histopathology images to map the ground truth cancer labels onto MRI. Shao *et al* (2021) proposed a deep learning model called ProsRegNet to simplify and accelerate MRI-histopathology image registration. The model achieved more accurate registration results that were more than 20 times faster than the traditional techniques.

Active surveillance: As highlighted earlier, AS is the only recommended management strategy for men with low-risk PCa (Mottet *et al* 2021). Men on AS are closely monitored with PSA measurements, repeat biopsies (only if needed), DRE, and mpMRI to detect disease progression (Shill *et al* 2021). Bozgo *et al* (2024) speculated whether MRI and AI can enhance the AS protocols. They suggested that AI could streamline the serial MRI assessments and decrease the inter-observer variability in reading the images (Sushentsev and Barrett 2022). There are two ways in which AI can achieve this—one is to help with more accurate characterization of prostate lesions so that the specificity of MRI can be enhanced (Oerther *et al* 2023). The second way is to monitor disease progression by analyzing sequential MRI scans and other clinical parameters. Bozgo *et al* (2024) provide a detailed summary of many studies that use ML and DL to monitor disease progression. They concluded that the performance of AI improved with the addition of clinicopathological characteristics, and changes between different time points were included in the modeling. More studies are needed to trigger an evidence-based change in the current AS protocols.

4.3.2 AI in transrectal ultrasound (TRUS)

TRUS is unsuitable as a standalone screening tool due to its high false-positive rate, but TRUS-guided prostate biopsy using a 12-core method seems to be a common practice (Lee and Chia 2015). However, even these systematic biopsies miss 40%–52% of csPCA (Ahdoot *et al* 2020, Ahmed *et al* 2017, Elkhoury *et al* 2019). Recently, Rusu *et al* (2025) trained ProCUSNet, based on nnUNet frameworks, to detect and outline csPCA on 3D reconstructions of 2D B-mode TRUS images. They reported that ProCUSNet found 82% csPCa and recommended it to be used as a supplement to systematic biopsies. Since it is comparable to MRI, they suggest that the algorithm can be used in the absence of MRI. More studies are needed in this area to utilize AI effectively in analyzing TRUS images, independently or with other clinical and imaging data.

4.3.3 AI in position emission tomography (PET)

Prostate-specific membrane antigen (PSMA) (a protein found on the surface of prostate cancer cells) expression is significantly upregulated in PCa. PET is a primary way to detect the presence of PSMA, and current international guidelines recommend a PSMA-PET scan for the primary staging of PCa prior to active treatment or to detect disease recurrence after active treatment (Schaeffer *et al* 2023). AI can assist in anatomical segmentation and PSMA uptake quantification tasks. Liu *et al* (2024) have reviewed several studies that developed AI algorithms for evaluating PSMA-PET scans and concluded that AI has the potential to detect metastasis, measure tumor burden, detect high-grade intraprostatic cancer, and predict treatment outcomes. Recently, the US FDA-approved DL-based AI software for evaluating PSMA-PET scans—Prostate Cancer Molecular Imaging Standardized Evaluation (aPROMISE) developed by EXINI Diagnostics AB (Lund, Sweden) (aPROMISE 2025).

4.3.4 AI in prostate pathology

Biopsies continue to be the gold standard for PCa confirmation. Traditionally, pathologists study tissue specimens and assign a Gleason grade. Gleason grade is a system used to classify and assess the aggressiveness of PCa by analyzing the microscopic appearance of cancer cells in the tissue sample. The scoring is from 1 to 5, where a score of 1 indicates well-differentiated cells consistent with normal prostate tissue, while a score of 5 indicates poorly-differentiated, more aggressive cells. The final Gleason score is calculated by adding the two most common grades of cancer cells found in the sample. A Gleason score of 6 or lower is generally considered low-risk, 7 and 8 are considered intermediate-risk, and a score of 8 or higher is considered high-risk. The Gleason grading method suffers from high inter-observer variability and is an area for potential AI support. Another area where AI has been extensively applied is to detect PCa automatically in digitized whole-slide images (WSIs) of prostate biopsy specimens.

Competitive challenges such as PANDA (Prostate cANcer graDe Assessment) have been organized to develop AI algorithms using the largest publicly available EU datasets of prostate biopsies (Bulten *et al* 2022). A key outcome was that they could fully reproduce top-performing algorithms and validate their generalization capabilities to independent US and EU cohorts. Such competitions are critical in every area of AI-based healthcare approaches to accelerate global research. 1920 developers participated in the challenge and developed reproducible AI algorithms for Gleason grading using 10 616 digitized prostate biopsies. Zhu *et al* (2024) reviewed the literature to highlight successful studies that used AI to detect and grade PCa, predict patient outcomes, identify molecular subtypes, and share challenges in developing AI for analyzing prostate pathology. They also list the various AI-based tools that recently received FDA clearance and CE certifications. One of them, which received FDA clearance, is Paige Prostate. More details of this tool are provided in section 4.6. Their article also includes a list of publicly available prostate pathology image datasets.

Multiple studies have shown great promise in using DL to detect cancer in WSI. Riaz *et al* (2024) present a detailed review of these studies, including those using AI for Gleason grading. Studies have also developed algorithms to identify known high-risk features on pathology images, such as cribriform patterns (Ambrosini *et al* 2020, Silva-Rodríguez *et al* 2020).

4.3.5 AI in risk stratification

Tests such as 4Kscore, PHI, PRS, Select MDx, and ExoDx Prostate Test, described in section 4.2, are typically used prior to a prostate biopsy to help assess the risk of csPCa and to avoid unwanted biopsies. Once the presence of csPCa is established via biopsy and a Gleason score is established, it is crucial to stratify risk and understand the aggressiveness of the tumor and/or the likelihood of developing aggressive cancer to guide the selection of the subsequent treatment strategy. Some of the most commonly used risk stratification tools used at this stage can be categorized into two: (1) Predictive models such as CAPRA (Cancer of the Prostate Risk Assessment) Score, Partin Tables, MSKCC (Memorial Sloan Kettering Cancer Center) nomograms that use clinical (Gleason score, PSA, tumor stage, etc) and biopsy data to determine if the cancer is localized or likely to spread. (2) Genome-based tests such as Oncotype DX GPS, Decipher, and Prolaris that perform genomic analysis of biopsy tissue to predict cancer aggressiveness and distinguish low-risk patients who can avoid treatment from high-risk patients. Decipher is also used after prostatectomy to assess recurrence risk and guide adjuvant therapy decisions. In a study by Zelic *et al* (2020), the prognostic performance of several of these approaches was compared. They found that the MSKCC nomogram, CAPRA score, and CPG (Cambridge Prognostic Groups) risk grouping program performed best in predicting prostate cancer death.

These conventional risk stratification methods are generally suboptimal and limited by inter-observer variabilities. AI-assisted risk stratification could address these issues, and several studies have demonstrated how ML and DL models can more accurately establish the risk of patients with localized prostate cancer (Riaz *et al* 2024). Most of these methods predicted the likelihood of biochemical recurrence (BCR) of cancer. Esteva *et al* (2022) developed a multimodal AI (MMAI) system that integrates clinical and digital histopathology data from prostate biopsies. The multimodal DL architecture demonstrated superior risk stratification performance (ranging from 9.2% to 14.6% relative improvement in a held-out validation set) compared to the tool developed by the National Cancer Center Network (NCCN). Shao *et al* (2024) proposed a DL model using clinico-pathologic data and digitized histopathology images. The model outperformed Gleason grading and CAPRA models by reclassifying 3.9% of low-risk patients as high-risk, and 21.3% of high-risk patients as low-risk.

Approximately one-third of the patients who undergo local therapy experience BCR. Detecting BCR is tricky as the PSA levels rise without objective evidence in imaging. In some patients with BCR, the disease can quickly progress to metastatic disease (Shore *et al* 2024). Therefore, an active area of research is to develop AI/ML approaches to identify patients at risk of disease progression and metastasis. Sabbagh *et al* (2024) developed an XGBoost model to predict the 5-year metastasis risk in patients with BCR. Another area is to select the optimal treatment strategy for these patients to delay or prevent metastasis.

4.3.6 AI in metastatic prostate cancer management

AI-powered solutions can help clinicians detect metastases earlier, optimize treatment strategies, and monitor treatment response. As highlighted earlier, PSMA-PET combined with AI has the potential to detect metastasis, measure tumor burden, and predict treatment outcomes (Erle *et al* 2021, Capobianco *et al* 2022, Lindgren Belal *et al* 2024). There are tools in the market, such as DeepPSMA, a software tool that utilizes AI technology to enable fast and reliable detection, staging, and quantification of prostate cancer using PSMA-PET/CT images (DeepPSMA 2025). Besides these applications, AI-based algorithms such as natural language processing (NLP) can be applied to clinical text (clinician notes, radiology reports, etc) from patients with metastatic cancer to predict disease progression and update cancer registries. LLMs can play a significant role in these areas. Riaz *et al* (2024) present a more in-depth review of how AI can improve prognostication and assist in clinical decision-making in managing advanced metastatic cancer.

4.3.7 AI-based classification of BPH and cancer

Benign prostatic hyperplasia (BPH), also known as enlarged prostate, is a non-cancerous condition where the prostate gland grows larger than usual. BPH and PCa often share similar symptoms, like urinary difficulties, and both conditions lead to elevated PSA levels in the blood. So, it is difficult to differentiate both conditions without a prostate biopsy more accurately. The enlarged prostate tissue from BPH can also mask the presence of cancerous lesions, requiring more targeting testing to detect cancer. BPH has a higher incidence rate than PCa (over 70% of men over 60 have BPH), so obtaining large datasets necessary to train AI models to advance detection and treatment decision-making is easy. Still, this is a growing area of research as most studies develop models for PCa and study BPH as an extension.

Shah *et al* (2022) reviewed AI models developed using data from 1600 patients from four studies. The models used fuzzy systems, computer-based vision systems, and clinical data mining to diagnose BPH, predict its severity, and determine the causative factors for treatment response. Fuzzy systems could predict BPH with 90% accuracy, and the vision systems detected it with 96.3% accuracy. The clinical data mining model predicted the clinical response to medical treatment with sensitivity and specificity of 70% and 50%, respectively. Bermejo *et al* (2015) proposed two predictive models to help clinicians differentiate BPH from PCa. The models used prostate volume, PSA levels, DRE results, and international prostate symptom scores to differentiate PC and BPH with an AUC of 72% and 80%. In a recent study by Huang *et al* (2023), DL models were developed to classify PCa and BPH in TRUS images. The model demonstrated a high classification performance with all metrics crossing 0.9400.

Recently, BPGbio, Inc. developed a PCa screening test that employed AI to analyze tissue and blood samples to differentiate PCa from BPH and rule out aggressive cancers. Using its Interrogative Biology platform to combine AI with a biobank of patient samples built over 12 years, BPGbio, Inc. detected biomarkers. The company developed a mass spectrometry-based blood test that detects a novel Filamin-A (FLNA) biomarker. They found that FLNA combined with prostate volume and age provided superior predictive performance to differentiate between men with BPH and PCa (Kiebish *et al* 2021). However, prostate volume, measured from TRUS images, could be subjective. So, the same group conducted another study to establish FLNA as an independent marker in differentiating BPH and PCa in two different ethnic populations (Mahaveer Chand *et al* 2024). McNally *et al* (2020) reviewed studies reporting biomarkers to differentiate BPH and cancer and found hundreds of potential biomarkers in urine, serum, tissue, and semen. However, more studies are needed to establish the utility of these biomarkers, and AI can be a powerful assistance tool for analysis.

4.4 Prostate cancer treatment planning

The treatment strategy for PCa depends on the PSA levels, cancer stage, Gleason score, metastasis risk, and the patient's overall health. Treatment options range from active surveillance and watchful waiting for low-risk cases to surgery, radiation, hormone therapy, chemotherapy, immunotherapy, and targeted therapies for high-risk cases. Active surveillance is generally recommended for low-risk localized cancer with a Gleason score $\leqslant 6$ and low PSA levels. The patients are offered regular PSA tests, DRE, and biopsies if needed during the waiting period. Active surveillance is generally recommended for older patients with other comorbid conditions where treatment may cause more harm than merit.

If the cancer is localized or locally advanced, surgery is usually recommended. Some surgery options are robot-assisted laparoscopic prostatectomy (RALP) and open radical prostatectomy. The most common side effects associated with such radical treatments are: erectile dysfunction (58%–79%), gastrointestinal toxicity and proctopathy (13%–34%), incontinence (range 3.2%–14%), overactive and/or obstructive urinary symptoms and hematuria (5%) (Resnick *et al* 2013, Corsini *et al* 2024). Because of these side effects, focal therapies (FT) are gaining acceptance as an alternative to prostatectomy and radiation therapy in cases where the tumor is confined to one area of the prostate (Connor *et al* 2020). Several types of FT exist, including cryotherapy, focal laser ablation (FLA), high-intensity focused ultrasound (HIFU), irreversible electroporation (IRE), transurethral ultrasound ablation of the prostate (TULSA).

One of the most popular FTs is FLA. It is a minimally invasive treatment option for localized prostate cancer. FLA is generally performed under local anesthesia or sedation, does not require large incisions, and targets and destroys cancerous tissue while sparing the surrounding healthy prostate tissue. As a result, it has fewer side effects than whole gland treatments. MRI or TRUS modalities are usually used to guide a laser fiber into the prostate to heat and destroy tumor cells. FLA is usually recommended for men with low-to intermediate-risk PCa with well-defined lesions that are localized to one or two areas of the prostate, identified with mpMRI. Some studies have concluded that FLA could have the highest risk of any-cause mortality but an insignificantly lower risk of cancer-specific mortality (Zheng *et al* 2019). The success of FT is highly dependent on accurate patient selection and disease localization. AI can help with more accurate tumor delineation than physicians, and several studies have established that (Priester *et al* 2023, Mota *et al* 2024). Unfold AI, a software cleared by the FDA uses an AI algorithm to generate 3D cancer estimation maps (CEM) to visualize cancer probability in 3D. The Unfold AI model was trained to generate 3D CEMs using multi-institutional, multi-modal input data consisting of MRI, prostate and MRI region of interest (ROI) segmentations, 3D biopsy locations, International Society of Urological Pathology Grade Group (GG), and PSA levels (Priester *et al* 2023).

Tătaru *et al* (2021) review studies using AI for treatment planning that involve radiotherapy. Some algorithms are commercially available. For example, ArteraAI's prostate test is an MMAI biomarker test that uses an AI algorithm to assess digitized biopsy images and clinical data to predict if the patient will benefit from hormone therapy (Spratt *et al* 2023). AI can also help in radiotherapy treatment planning by assisting in dose distribution calculations (Church *et al* 2024).

Despite the availability of a variety of therapeutic options for PCa, therapeutic resistance, disease recurrence, and metastasis are still prevalent challenges. Derbal (2025) presented an interesting hypothetical case study on how LLMs could treat metastatic PCa via generative pre-trained transformers (GPTs)-supported adaptive intermittent therapy. As LLM research evolves, we can expect more such applications.

4.5 LLM applications

Risk forecasting: Foundation models are now being exploited to develop real-world risk forecasting of diseases. Electronic health records (EHRs) contain detailed longitudinal information about a patient's clinical history, stored in both structured (demographics, laboratory results, medication lists, or diagnosis codes) and unstructured free text (doctor's notes, imaging reports, and other correspondences). Forecasting is a process of predicting future clinical events based on historical data and is generally done using structured data. However, it is missing information stored in unstructured data, constituting about 80% of patient data. Kraljevic *et al* (2024) recently proposed Foresight, a generative pre-trained transformer-based pipeline that uses entity recognition and linking tools to convert unstructured EHR data into structured, coded concepts. Then, a DL model was built to forecast disorders and biomedical concepts more generally. The study team has also made the model publicly available via a web application (https://foresight.sites.er.kcl.ac.uk/) (figure 4.1). Such models show promise in the temporal modeling of large-scale longitudinal individual patient data to do risk forecasting.

Clinical trials: AI platforms could also ideally match metastatic prostate cancer patients with suitable clinical trials based on their genomic profile, disease stage, and past treatments. Generic LLMs like TrialGPT are being developed and tested (Jin *et al* 2024). It was observed that TrialGPT could reduce patient recruitment screening time by 42.6%. Clinical trials are costly and time-consuming. LLMs are also being studied to accelerate evidence generation by generating synthetic patient data (Eckardt *et al* 2024).

Patient education: Alasker *et al* (2024) evaluated the performance of LLMs (ChatGPT-3.5, ChatGPT-4, and Bard) in providing patient education on PCa and concluded that they could do so. It is still a hugely underresearched area as LLMs are not yet skilled in analyzing questions in specific contexts and cannot ask questions back to gather more context (Zhu *et al* 2023).

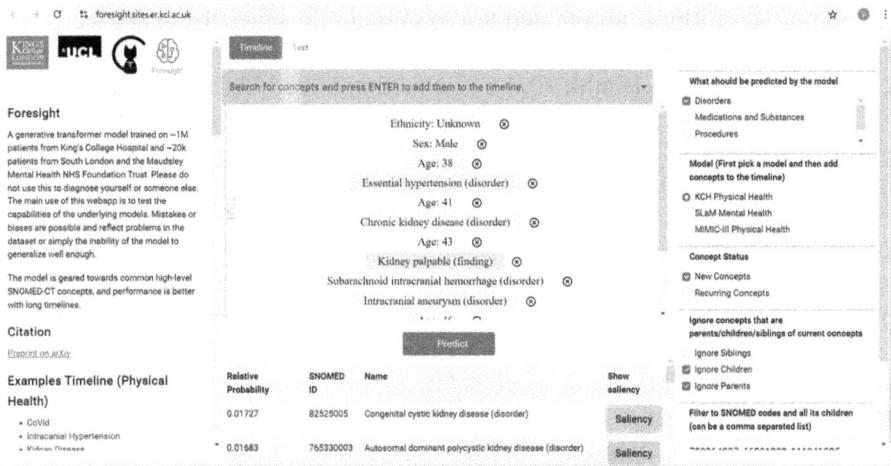

Figure 4.1. Screenshot of the Foresight web application that uses foundation models to use structured and unstructured data from a patient's EHR to forecast the risk of disorders, offer differential diagnoses, suggest substances be used, etc. Image credit: Foresight web application hosted by King's College London.

4.6 FDA/CE-approved commercial AI applications

Regulatory clearance is necessary for CAD solutions to be used in clinical practice. In the United States, the Food and Drug Administration (FDA) is responsible for evaluating the safety and efficacy of these solutions; in Europe, a CE mark is necessary. To date, the FDA has approved over 1000 (1016 to be exact) medical solutions that incorporate AI and ML (FDA AI/ML Devices 2025). 777 of these 1016 solutions (76.5%) are in the field of radiology. Ten devices have been approved to date for prostate cancer management (table 4.1).

MEDIHUB PROSTATE (JLK 2025): JLK, a South Korea-based AI company, developed MEDIHUB, an AI-powered medical imaging analysis platform for a suite of diseases. The prostate module helps urologists and radiologists analyze multiparametric MRI images using AI to outline the prostate semi-automatically. It also provides visualizations and structured reports to support clinical decision-making. In June 2024, MEDIHUB Prostate received FDA 510(k) clearance, marking a significant milestone in its validation and adoption. It also has received CE marking.

Quantib® Prostate (Quantib 2025): Quantib, now a subsidiary of RadNet, Inc., has received multiple FDA clearances and CE markings for its AI-powered product for prostate MRI management. In October 2020, Quantib received FDA clearance for its initial AI prostate solution, Quantib® Prostate, which is considered the first comprehensive AI prostate software available to radiologists in the United States. In May 2022, the FDA-cleared Quantib® Prostate 2.0, which offers tools to improve reporting quality and speed, including AI-based segmentation, PSA density calculation, precise registration and movement correction, one-click segmentation of lesion candidates, PI-RADS scoring support, standardized reporting. In April 2023, the FDA-cleared Quantib® Prostate 3.0 for its utility in enhancing prostate MRI reading and reporting. This version included improved prostate and sub-region segmentation algorithms, automated lesion mapping on the PI-RADS sector map, and other user experience enhancements.

Unfold AI™ (Avenda Health 2025): In December 2022, Avenda Health's Unfold AI™ received 510(k) clearance from the FDA. The AI-powered clinical decision support software uses PSA, MRI, biopsy, and pathology data to generate a cancer estimation map (CEM). It is designed to be used alongside MRI and biopsy results to enhance diagnostic accuracy and treatment planning. The decision support software, the first of its kind, helps in tumor characterization (identification of the size and extent of the disease), treatment selection, treatment guidance, and improved patient understanding of the disease and the treatment. One of their clinical studies indicated that about 28% of urologists change their treatment plans after using Unfold AI (Brisbane *et al* 2023).

ProstatID™ (Bot Image 2025): ProstatID™ is the first and only FDA-cleared AI software for screening, detecting, and diagnosing PCa in MRI images with about 93.6% AUC. It is a software interface for post-processing MRI images using cloud-based computing and AI image interpretation. The system also uses PI-RADS to score the cancer probability of suspicious lesions. The software features include

Table 4.1. FDA-approved (510(k) cleared) devices with AI/ML for prostate cancer management.

Date of final decision	Submission number	Device	Company	Key AI features
06/21/2024	K233196	MEDIHUB PROSTATE	JLK Inc.	AI-based multi-sequential MRI analysis to detect suspected prostate cancer areas
04/17/2023	K230772	Quantib® Prostate 3.0	Quantib B.V.	AI-driven prostate segmentation, volume measurements, and integrated automatic PSA density calculation; Visual PI-RADS v.2.1 scoring support (MRI)
11/22/2022	K221624	Unfold AI™ AI Prostate Cancer Planning Software	Avenda Health, Inc.	AI-powered clinical decision support software to generate a cancer estimation map (PSA, MRI, biopsy, and pathology data)
07/08/2022	K212783	ProstatID™	Bot Image™	AI-based automatic suspicious lesion detection, segmentation, sizing, classification, diagnosis, treatment planning and guidance (MRI)
05/13/2022	K221106	Quantib® Prostate 2.0	Quantib B.V.	Improved AI-based segmentation, PSA density calculation, precise registration and movement correction, PI-RADS scoring support, and standardized reporting (MRI)
09/21/2021	DEN200080	Paige Prostate	Paige.AI	The first AI-based FDA-cleared pathology for cancer detection in prostate needle biopsies (Pathology)
02/04/2021	K203582	QP-Prostate®	Quibim	Automated segmentation of the prostate gland and seminal vesicles, PSA density calculations, fusion biopsy planning, AI-driven cancerous lesion identification (MRI)
11/17/2020	K193306	PROView	GE Healthcare	Automated prostate segmentation for volume and PSADcalculation, workflow optimization (MRI)
10/11/2020	K202501	Quantib® Prostate	Quantib BV	AI-based volumetry and PSA density, registration and movement correction, fast segmentation, workflow optimization (MRI)
07/30/2020	K193283	AI-Rad Companion Prostate MR	Siemens Healthineers	MRI reading and reporting and automatic prostate segmentation for targeted MR/US-fusion biopsies

automatic suspicious lesion detection, segmentation, sizing, classification, diagnosis, treatment planning, and guidance using a 3D cognitive targeting system.

Paige Prostate (Paige.AI 2025, Paige Prostate Suite 2024): Paige Prostate is an FDA-approved AI software developed by Paige.AI to assist pathologists in detecting prostate cancer. It analyzes digitized whole-slide images of prostate needle biopsies to identify suspicious areas and provide the coordinates of the region with the highest likelihood of cancer. The FDA clearance was obtained based on a study where 16 pathologists studied 527 digitized slide images (171 cancerous and 356 benign), both unassisted and assisted by the software. Cancer detection improved by 7.3% when using the software.

QP-Prostate® (Quibim 2025): QP-Prostate® is AI-powered FDA-cleared software that enhances the analysis of prostate MRI scans. The software offers automated segmentation of the prostate gland and seminal vesicles, PSA density calculations, fusion biopsy planning, and AI-driven cancerous lesion identification. In October 2023, Quibim announced a collaboration with Philips to integrate QP-Prostate® into Philips' AI-enabled MR imaging systems.

PROView (GE Healthcare 2025): PROView is an advanced FDA-cleared prostate MRI image analysis software. It performs automated prostate segmentation using deep learning algorithms and aids in assessing prostate gland volume.

AI-Rad Companion Prostate MR (Siemens Healthineers 2025): This digital companion AI software that is both FDA-cleared and CE-certified supports prostate MRI reading and reporting and annotating prostate gland images for targeted MR/US-fusion biopsies.

Other CE-approved solutions: Lucida Medical's Prostate Intelligence™ (Pi™) is a CE-certified AI-driven software to analyze MRI scans to detect prostate cancer (Lucida Medical 2025). Ibex Medical Analytics' Galen™ Prostate (Pantanowitz *et al* 2020, IBEX-AI 2025) is another CE-registered AI-powered diagnostic solution that assists pathologists in the primary diagnosis of prostate biopsies. Indica Labs' HALO Prostate AI (Indica Labs 2025) is a deep learning-based CE-certified screening tool designed to assist pathologists in identifying and grading prostate cancer in core needle biopsies.

4.7 Conclusions

Prostate cancer management has been well supported by advancements in AI, as described in this chapter. Researchers worldwide have conducted several studies to develop AI-supported solutions for PCa, but there has been limited progress in translating this research to clinical settings or commercial use. Fragmented research by different groups will impede progress, but that is the norm currently. To accelerate this progress, more efforts are needed to enable strong collaborations among research groups, data-rich hospitals, healthcare providers, regulators, and industry stakeholders. Besides these aspects, developing an AI solution always comes with standard challenges—the lack of large-scale, well-curated data from relevant patient cohorts is a big one. AI models are data-hungry, but as the adage goes, 'garbage in, garbage out.' For AI models to successfully analyze MRI images, for example, digital annotations of the images, histopathological correlation with tissue samples, and tumor mapping for MRI verification are key. However, these are not done in routine clinical procedures in most hospitals as they are costly and time-intensive processes that the already busy physicians do not have time for. There is also no established reference standard for annotations, and the accuracy is subjective and dependent on human readers whose experience and expertise vary.

Considering the AI models, they are susceptible to overfitting and bias. Most studies use single institution small cohorts. Models developed using such screened small-scale studies do not generalize well enough to be similarly accurate in populations with different risk levels, from different institutions using different imaging modalities, and from different demographics. All these limitations could be why we still observe a high false-positive rate in PCa diagnosis (Penzkofer *et al* 2021). Balancing sensitivity and the false-positive rate is critical, and well-designed, large-scale studies could ideally help. Other issues are patient data privacy and security concerns, clinician adoption due to interoperability issues, lack of algorithm explainability and transparency, lack of education, etc (Agrawal and Vagha 2024). The long-term success of AI in prostate cancer depends on how various stake-holders, including patients, come together to solve multiple ethical, legal, regulatory, and technical concerns.

References

Adams L C, Makowski M R, Engel G *et al* 2022 Prostate158—an expert-annotated 3T MRI dataset and algorithm for prostate cancer detection *Comput. Biol. Med.* **148** 105817

Agrawal S and Vagha S 2024 A comprehensive review of artificial intelligence in prostate cancer care: state-of-the-art diagnostic tools and future outlook *Cureus* **16** e66225

Ahdoot M, Wilbur A R, Reese S E *et al* 2020 MRI-targeted, systematic, and combined biopsy for prostate cancer diagnosis *N. Engl. J. Med.* **382** 917–28

Ahmed H U, El-Shater Bosaily A, Brown L C *et al* 2017 Diagnostic accuracy of multi-parametric MRI and TRUS biopsy in prostate cancer (PROMIS): a paired validating confirmatory study *Lancet* **389** 815–22

Alasker A, Alsalamah S, Alshathri N *et al* 2024 Performance of large language models (LLMs) in providing prostate cancer information *BMC Urol* **24** 177

Alford A V, Brito J M, Yadav K K, Yadav S S, Tewari A K and Renzulli J 2017 The use of biomarkers in prostate cancer screening and treatment *Rev. Urol* **19** 221–34

Ambrosini P, Hollemans E, Kweldam C F *et al* 2020 Automated detection of cribriform growth patterns in prostate histology images *Sci. Rep.* **10** 14904

aPROMISE 2025 https://lantheusholdings.gcs-web.com/news-releases/news-release-details/lan-theus-ai-enabled-apromise-now-available-siemens-healthineers

Armato S G, Huisman H, Drukker K *et al* 2018 PROSTATEx challenges for computerized classification of prostate lesions from multiparametric magnetic resonance images *J. Med. Imaging (Bellingham)* **5** 044501

Artan Y and Yetik I S 2012 Prostate cancer localization using multiparametric MRI based on semisupervised techniques with automated seed initialization *IEEE Trans. Inf. Technol. Biomed.* **16** 1313–23

Avenda Health 2025 https://avendahealth.com/unfold-ai-based-decision-support/

Baboudjian M, Rouprêt M and Ploussard G 2023 Redefining Gleason 6 prostate cancer nomenclature: the surgeon's perspective *J. Clin. Oncol* **41** 1492–3

Bermejo P, Vivo A, Tárraga P J and Rodríguez-Montes J A 2015 Development of interpretable predictive models for BPH and prostate cancer *Clin. Med. Insights Oncol* **9** 15–24

Bhandary S, Kuhn D, Babaiee Z *et al* 2023 Investigation and benchmarking of U-Nets on prostate segmentation tasks *Comput. Med. Imaging Graph.* **107** 102241

Bozgo V, Roest C, van Oort I, Yakar D, Huisman H and de Rooij M 2024 Prostate MRI and artificial intelligence during active surveillance: should we jump on the bandwagon? *Eur. Radiol* **34** 7698–704

Bray F, Laversanne M, Sung H *et al* 2024 Global cancer statistics 2022: GLOBOCAN estimates of incidence and mortality worldwide for 36 cancers in 185 countries *CA Cancer J. Clin* **74** 229–63

Brisbane W, Priester A, Mota S M, Shubert J, Bong J, Sayre J *et al* 2023 MP73-12 artificial intelligence-based cancer mapping to aid in prostate cancer therapy *J. Urol* **209** e1039

Bot Image 2025 https://botimageai.com/prostatid/

Bulten W, Kartasalo K, Chen P C *et al* 2022 Artificial intelligence for diagnosis and Gleason grading of prostate cancer: the PANDA challenge *Nat. Med.* **28** 154–63

Capobianco N, Sibille L, Chantadisai M *et al* 2022 Whole-body uptake classification and prostate cancer staging in 68Ga-PSMA-11 PET/CT using dual-tracer learning *Eur. J. Nucl. Med. Mol. Imaging* **49** 517–26

Chen S, Jian T, Chi C *et al* 2022 Machine learning-based models enhance the prediction of prostate cancer *Front. Oncol* **12** 941349

Chen T, Hu W, Zhang Y *et al* 2025 A multimodal deep learning nomogram for the identification of clinically significant prostate cancer in patients with gray-zone PSA levels: comparison with clinical and radiomics models *Acad. Radiol.* **32** 864–76

Church C, Yap M, Bessrour M, Lamey M and Granville D 2024 Automated plan generation for prostate radiotherapy patients using deep learning and scripted optimization *Phys. Imaging Radiat. Oncol* **32** 100641

Connor M J, Gorin M A, Ahmed H U and Nigam R 2020 Focal therapy for localized prostate cancer in the era of routine multi-parametric MRI *Prostate Cancer Prostatic Dis* **23** 232–43

Corsini C, Bergengren O, Carlsson S *et al* 2024 Patient-reported side effects 1 year after radical prostatectomy or radiotherapy for prostate cancer: a register-based nationwide study *Eur. Urol. Oncol* **7** 605–13

David M K and Leslie S W 2024 Prostate-specific antigen *StatPearls* (Treasure Island, FL: StatPearls Publishing)

DeepPSMA 2025 https://nucs.ai/products/deep-psma

Derbal Y 2025 Adaptive treatment of metastatic prostate cancer using generative artificial intelligence *Clin. Med. Insights Oncol.* **19** 11795549241311408

Drost F H, Osses D F, Nieboer D *et al* 2019 Prostate MRI, with or without MRI-targeted biopsy, and systematic biopsy for detecting prostate cancer *Cochrane Database Syst. Rev.* **4** CD012663

Dunn B K, Woloshin S, Xie H and Kramer B S 2022 Cancer overdiagnosis: a challenge in the era of screening *J. Natl Cancer Cent* **2** 235–42

Eckardt J N, Hahn W, Röllig C *et al* 2024 Mimicking clinical trials with synthetic acute myeloid leukemia patients using generative artificial intelligence *NPJ Digit. Med* **7** 76

Eggener S E, Berlin A, Vickers A J, Paner G P, Wolinsky H and Cooperberg M 2022 Low-grade prostate cancer: time to stop calling it cancer *J. Clin. Oncol* **40** 3110–4

Elkhoury F F, Felker E R, Kwan L *et al* 2019 Comparison of targeted vs systematic prostate biopsy in men who are biopsy naive: the prospective assessment of image registration in the diagnosis of prostate cancer (PAIREDCAP) study *JAMA Surg* **154** 811–8

Epstein J I and Kibel A S 2022 Renaming gleason score 6 prostate to noncancer: a flawed idea scientifically and for patient care *J. Clin. Oncol* **40** 3106–9

Erle A, Moazemi S, Lütje S, Essler M, Schultz T and Bundschuh R A 2021 Evaluating a machine learning tool for the classification of pathological uptake in whole-body PSMA-PET-CT scans *Tomography* **7** 301–12

Esserman L J, Thompson I M Jr and Reid B 2013 Overdiagnosis and overtreatment in cancer: an opportunity for improvement *JAMA* **310** 797–98

Esserman L J, Thompson I M, Reid B *et al* 2014 Addressing overdiagnosis and overtreatment in cancer: a prescription for change *Lancet Oncol.* **15** e234–42

Esteva A, Feng J, van der Wal D *et al* 2022 Prostate cancer therapy personalization via multi-modal deep learning on randomized phase III clinical trials [published correction appears in *NPJ Digit. Med.* 2023 Feb 22;6(1):27. 10.1038/s41746-023-00769-z] *NPJ Digit. Med.* **5** 71

FDA AI/ML Devices 2025 https://fda.gov/medical-devices/software-medical-device-samd/artificial-intelligence-and-machine-learning-aiml-enabled-medical-devices

GE Healthcare 2025 https://gehealthcare.com/en-sg/products/magnetic-resonance-imaging/pro-view-body?srsltid=AfmBOooW3iMINru4wi91RyGL1_Wlgt1YqB4eUczkJmR8IAaguDde6d2E

Giganti F, Allen C, Emberton M, Moore C M and Kasivisvanathan V 2020 PRECISION study group. Prostate imaging quality (PI-QUAL): a new quality control scoring system for multiparametric magnetic resonance imaging of the prostate from the PRECISION trial *Eur. Urol. Oncol* **3** 615–9

Giganti F, Lindner S, Piper J W *et al* 2021 Multiparametric prostate MRI quality assessment using a semi-automated PI-QUAL software program *Eur. Radiol. Exp* **5** 48

Gong E, Pauly J M, Wintermark M and Zaharchuk G 2018 Deep learning enables reduced gadolinium dose for contrast-enhanced brain MRI *J. Magn. Reson. Imaging* **48** 330–40

Griswold M A, Jakob P M, Heidemann R M *et al* 2002 Generalized autocalibrating partially parallel acquisitions (GRAPPA) *Magn. Reson. Med.* **47** 1202–10

Hong Y, Qiu Z, Chen H, Zhu B and Lei H 2023 MAS-UNet: a U-shaped network for prostate segmentation *Front. Med. (Lausanne)* **10** 1190659

Huang T L, Lu N H, Huang Y H *et al* 2023 Transfer learning with CNNs for efficient prostate cancer and BPH detection in transrectal ultrasound images *Sci. Rep.* **13** 21849

IBEX-AI 2025 https://ibex-ai.com/prostate/

Indica Labs 2025 https://indicalab.com/news/halo-prostate-ai-a-tool-for-automated-detection-and-gleason-grading-of-prostate-cancer/

Islami F, Ward E M, Sung H *et al* 2021 Annual report to the nation on the status of cancer, part 1: national cancer statistics *J. Natl Cancer Inst* **113** 1648–69

Jin Q, Wang Z, Floudas C S *et al* 2024 Matching patients to clinical trials with large language models *Nat. Commun.* **15** 9074

JLK 2025 https://jlkgroup.com/en/medihub_products/

Johnson P M, Tong A, Donthireddy A *et al* 2022 Deep learning reconstruction enables highly accelerated biparametric MR imaging of the prostate *J. Magn. Reson. Imaging* **56** 184–95

Kiebish M A, Tekumalla P, Ravipaty S *et al* 2021 Clinical utility of a serum biomarker panel in distinguishing prostate cancer from benign prostate hyperplasia *Sci. Rep.* **11** 15052

Kim S H, Kim S, Joung J Y *et al* 2018 Lifestyle risk prediction model for prostate cancer in a Korean population *Cancer Res. Treat* **50** 1194–202

Kim H, Kang S W, Kim J H *et al* 2023 The role of AI in prostate MRI quality and interpretation: opportunities and challenges [published correction appears in *Eur. J. Radiol.* 2024 Aug;177:111585. 10.1016/j.ejrad.2024.111585] *Eur. J. Radiol.* **165** 110887

Kraljevic Z, Bean D, Shek A *et al* 2024 Foresight-a generative pretrained transformer for modelling of patient timelines using electronic health records: a retrospective modelling study [published correction appears in *Lancet Digit. Health* 2024;6:e281–90] *Lancet Digit. Health* **6** e680

Labbate C V, Paner G P and Eggener S E 2022 Should grade group 1 (GG1) be called cancer? *World J. Urol* **40** 15–9

Leal J, Welton N J, Martin R M *et al* 2018 Estimating the sensitivity of a prostate cancer screening programme for different PSA cut-off levels: a UK case study *Cancer Epidemiol* **52** 99–105

Lee A and Chia S J 2015 Contemporary outcomes in the detection of prostate cancer using transrectal ultrasound-guided 12-core biopsy in Singaporean men with elevated prostate specific antigen and/or abnormal digital rectal examination *Asian J. Urol* **2** 187–93

Li H, Lee C H, Chia D, Lin Z, Huang W and Tan C H 2022 Machine learning in prostate MRI for prostate cancer: current status and future opportunities *Diagnostics (Basel)* **12** 289

Lindgren Belal S, Frantz S, Minarik D *et al* 2024 Applications of artificial intelligence in PSMA PET/CT for prostate cancer imaging *Semin. Nucl. Med.* **54** 141–9

Liu J, Sandhu K, Woon D T S, Perera M and Lawrentschuk N 2024 The value of artificial intelligence in prostate-specific membrane antigen positron emission tomography: an update *Semin. Nucl. Med.* https://doi.org/10.1053/j.semnuclmed.2024.12.001

Loeb S, Bjurlin M A, Nicholson J *et al* 2014 Overdiagnosis and overtreatment of prostate cancer *Eur. Urol* **65** 1046–55

Lucida Medical 2025 https://lucidamedical.com/pi/

Mahaveer Chand N, Tekumalla P K, Rosenberg M T *et al* 2024 Filamin A is a prognostic serum biomarker for differentiating benign prostatic hyperplasia from prostate cancer in Caucasian and African American men *Cancers (Basel)* **16** 712

McCaffery K, Nickel B, Pickles K *et al* 2019 Resisting recommended treatment for prostate cancer: a qualitative analysis of the lived experience of possible overdiagnosis *BMJ Open* **9** e026960

McNally C J, Ruddock M W, Moore T and McKenna D J 2020 Biomarkers that differentiate benign prostatic hyperplasia from prostate cancer: a literature review *Cancer Manag. Res* **12** 5225–41

Mota S M, Priester A, Shubert J *et al* 2024 Artificial intelligence improves the ability of physicians to identify prostate cancer extent *J. Urol* **212** 52–62

Mottet N, van den Bergh R C N, Briers E *et al* 2021 EAU-EANM-ESTRO-ESUR-SIOG guidelines on prostate cancer-2020 update. Part 1: screening, diagnosis, and local treatment with curative intent *Eur. Urol* **79** 243–62

Nitta S, Tsutsumi M, Sakka S *et al* 2019 Machine learning methods can more efficiently predict prostate cancer compared with prostate-specific antigen density and prostate-specific antigen velocity *Prostate Int* **7** 114–8

Oerther B, Engel H, Nedelcu A *et al* 2023 Prediction of upgrade to clinically significant prostate cancer in patients under active surveillance: performance of a fully automated AI-algorithm for lesion detection and classification *Prostate* **83** 871–8

Ozer S, Langer D L, Liu X, Haider M A, van der Kwast T H, Evans A J, Yang Y, Wernick M N and Yetik I S 2010 Supervised and unsupervised methods for prostate cancer segmentation with multispectral MRI *Med. Phys.* **37** 1873–83 PMID: 20443509

Paige.AI 2025 https://info.paige.ai/prostate

Paige Prostate Suite 2024 *The Paige Prostate Suite: Assistive Artificial Intelligence for Prostate Cancer Diagnosis: Emerging Health Technologies* (Ottawa, ON: Canadian Agency for Drugs and Technologies in Health) https://ncbi.nlm.nih.gov/books/NBK608438/

Pal A and Rathi Y 2022 A review and experimental evaluation of deep learning methods for MRI reconstruction *J. Mach. Learn. Biomed. Imaging* **1** 001

Pantanowitz L, Quiroga-Garza G M, Bien L *et al* 2020 An artificial intelligence algorithm for prostate cancer diagnosis in whole slide images of core needle biopsies: a blinded clinical validation and deployment study *Lancet Digit. Health* **2** e407–16

Penzkofer T, Padhani A R, Turkbey B *et al* 2021 ESUR/ESUI position paper: developing artificial intelligence for precision diagnosis of prostate cancer using magnetic resonance imaging *Eur. Radiol* **31** 9567–78

Perera M, Mirchandani R, Papa N *et al* 2021 PSA-based machine learning model improves prostate cancer risk stratification in a screening population *World J. Urol.* **39** 1897–902

Priester A, Fan R E, Shubert J *et al* 2023 Prediction and mapping of intraprostatic tumor extent with artificial intelligence *Eur. Urol. Open Sci.* **54** 20–7

Pruessmann K P, Weiger M, Scheidegger M B and Boesiger P 1999 SENSE: sensitivity encoding for fast MRI *Magn. Reson. Med.* **42** 952–62

Quantib 2025 https://quantib.com/en/solutions/quantib-prostate

Quibim 2025 https://quibim.com/qp-prostate/

Resnick M J, Koyama T, Fan K H *et al* 2013 Long-term functional outcomes after treatment for localized prostate cancer *N. Engl. J. Med.* **368** 436–45

Riaz I B, Harmon S, Chen Z, Naqvi S A A and Cheng L 2024 Applications of artificial intelligence in prostate cancer care: a path to enhanced efficiency and outcomes *Am. Soc. Clin. Oncol. Educ. Book* **44** e438516

Roobol M J and Carlsson S V 2013 Risk stratification in prostate cancer screening [published correction appears in *Nat. Rev. Urol.* 2013 May;10(5):248] *Nat. Rev. Urol.* **10** 38–48

Rosenkrantz A B, Verma S, Choyke P *et al* 2016 Prostate magnetic resonance imaging and magnetic resonance imaging targeted biopsy in patients with a prior negative biopsy: a consensus statement by AUA and SAR *J. Urol* **196** 1613–8

Rusu M, Jahanandish H, Vesal S *et al* 2025 ProCUSNet: prostate cancer detection on B-mode transrectal ultrasound using artificial intelligence for targeting during prostate biopsies *Eur. Urol. Oncol* **8** 477–85

Sabbagh A, Tilki D, Feng J *et al* 2024 Multi-institutional development and external validation of a machine learning model for the prediction of distant metastasis in patients treated by salvage radiotherapy for biochemical failure after radical prostatectomy *Eur. Urol. Focus* **10** 66–74

Saha A *et al* 2022 The PI-CAI challenge: public training and development dataset *v2.0, Zenodo* https://zenodo.org/records/6624726

Saha A, Bosma J S, Twilt J J *et al* 2024 Artificial intelligence and radiologists in prostate cancer detection on MRI (PI-CAI): an international, paired, non-inferiority, confirmatory study *Lancet Oncol.* **25** 879–87

Schaeffer E, Srinivas S, Antonarakis E S *et al* 2021 NCCN guidelines insights: prostate cancer, version 1.2021 *J. Natl Compr. Canc. Netw* **19** 134–43

Schaeffer E M, Srinivas S, Adra N *et al* 2023 Prostate cancer, version 4.2023, NCCN clinical practice guidelines in oncology *J. Natl Compr. Canc. Netw* **21** 1067–96

Shah M, Naik N, Hameed B Z *et al* 2022 Current applications of artificial intelligence in benign prostatic hyperplasia *Turk. J. Urol* **48** 262–7

Shao W, Banh L, Kunder C A *et al* 2021 ProsRegNet: a deep learning framework for registration of MRI and histopathology images of the prostate *Med. Image Anal.* **68** 101919

Shao Y, Bazargani R, Karimi D *et al* 2024 Prostate cancer risk stratification by digital histopathology and deep learning *JCO Clin. Cancer Inform.* **8** e2300184

Shen Q, Zheng B, Li W *et al* 2025 MixUNETR: a U-shaped network based on W-MSA and depth-wise convolution with channel and spatial interactions for zonal prostate segmentation in MRI *Neural Netw.* **181** 106782

Shill D K, Roobol M J, Ehdaie B, Vickers A J and Carlsson S V 2021 Active surveillance for prostate cancer *Transl. Androl. Urol* **10** 2809–19

Shore N D, Moul J W, Pienta K J, Czernin J, King M T and Freedland S J 2024 Biochemical recurrence in patients with prostate cancer after primary definitive therapy: treatment based on risk stratification *Prostate Cancer Prostatic Dis* **27** 192–201

Siemens Healthineers 2025 https://siemens-healthineers.com/en-th/digital-health-solutions/digital-solutions-overview/clinical-decision-support/ai-rad-companion/prostate-Mr

Silva-Rodríguez J, Colomer A, Sales M A *et al* 2020 Going deeper through the Gleason scoring scale: an automatic end-to-end system for histology prostate grading and cribriform pattern detection *Comput. Methods Programs Biomed.* **195** 105637

Smith C P, Harmon S A, Barrett T *et al* 2019 Intra- and interreader reproducibility of PI-RADSv2: a multireader study *J. Magn. Reson. Imaging* **49** 1694–703

Spratt D E, Tang S, Sun Y *et al* 2023 Artificial intelligence predictive model for hormone therapy use in prostate cancer *NEJM Evid* **2** EVIDoa2300023

Sushentsev N and Barrett T 2022 The role of artificial intelligence in MRI-driven active surveillance in prostate cancer *Nat. Rev. Urol.* **19** 510´

Tătaru O S, Vartolomei M D, Rassweiler J J *et al* 2021 Artificial intelligence and machine learning in prostate cancer patient management-current trends and future perspectives *Diagnostics (Basel)* **11** 354

Tong K, Wu Y and Zhou F 2020 Recent advances in small object detection based on deep learning: a review *Image Vis. Comput* **97** 103910

Tran T T, Lee J, Kim J, Kim S Y, Cho H and Kim J 2024 Machine learning algorithms that predict the risk of prostate cancer based on metabolic syndrome and sociodemographic characteristics: a prospective cohort study *BMC Public Health* **24** 3549

Turkbey B, Rosenkrantz A B, Haider M A *et al* 2019 Prostate imaging reporting and data system version 2.1: 2019 update of prostate imaging reporting and data system version 2 *Eur. Urol* **76** 340–51

Twilt J J, van Leeuwen K G, Huisman H J, Fütterer J J and de Rooij M 2021 Artificial intelligence based algorithms for prostate cancer classification and detection on magnetic resonance imaging: a narrative review *Diagnostics (Basel)* **11** 959 van der Leest M, Cornel E, Israël B *et al* 2019 Head-to-head comparison of transrectal ultrasound-guided prostate biopsy versus multiparametric prostate resonance imaging with subsequent magnetic resonance-guided biopsy in biopsy-naïve men with elevated prostate-specific antigen: a large prospective multicenter clinical study *Eur. Urol* **75** 570–8

Wang L, Lu B, He M, Wang Y, Wang Z and Du L 2022 Prostate cancer incidence and mortality: global status and temporal trends in 89 countries from 2000 to 2019 *Front. Public Health* **10** 811044

WCRF 2025 https://wcrf.org/preventing-cancer/cancer-types/prostate-cancer/

Welch H G and Albertsen P C 2009 Prostate cancer diagnosis and treatment after the introduction of prostate-specific antigen screening: 1986–2005 *J. Natl Cancer Inst* **101** 1325–9

Wei J T, Barocas D, Carlsson S *et al* 2023a Early detection of prostate cancer: AUA/SUO guideline part I: prostate cancer screening *J. Urol* **210** 45–53

Wei J T, Barocas D, Carlsson S *et al* 2023b Early detection of prostate cancer: AUA/SUO guideline part II: considerations for a prostate biopsy *J. Urol* **210** 54–63

Wu M, He X, Li F, Zhu J, Wang S and Burstein P D 2023 Weakly supervised volumetric prostate registration for MRI-TRUS image driven by signed distance map *Comput. Biol. Med.* **163** 107150

Xie H, Lei Y, Wang T *et al* 2022 Magnetic resonance imaging contrast enhancement synthesis using cascade networks with local supervision *Med. Phys.* **49** 3278–87

Yeo Y, Shin D W, Lee J *et al* 2021 Personalized 5-year prostate cancer risk prediction model in Korea based on nationwide representative data *J. Pers. Med* **12** 2

Zelic R, Garmo H, Zugna D *et al* 2020 Predicting prostate cancer death with different pretreatment risk stratification tools: a head-to-head comparison in a nationwide cohort study [published correction appears in *Eur. Urol.* 2020 Jul;78(1):e45–e47. 10.1016/j.eururo.2020.03.016] *Eur. Urol.* **77** 180–8

Zheng X, Jin K, Qiu S *et al* 2019 Focal laser ablation versus radical prostatectomy for localized prostate cancer: survival outcomes from a matched cohort *Clin. Genitourin. Cancer* **17** 464–469.e3

Zhu L, Mou W and Chen R 2023 Can the ChatGPT and other large language models with internet-connected database solve the questions and concerns of patient with prostate cancer and help democratize medical knowledge? *J. Transl. Med* **21** 269

Zhu L, Pan J, Mou W *et al* 2024 Harnessing artificial intelligence for prostate cancer management *Cell Rep. Med* **5** 101506

Ziglioli F, Patera A, Isgrò G, Campobasso D, Guarino G and Maestroni U 2023 Impact of modifiable lifestyle risk factors for prostate cancer prevention: a review of the literature *Front. Oncol* **13** 1203791

IOP Publishing

Predictive Analytics in Healthcare, Volume 2
Transforming the future of medicine
Vinithasree Subbhuraam

Chapter 5

How is AI transforming precision medicine?

Precision medicine is a personalized healthcare model that uses computational science, genetics, biology, and social factors to understand and prevent disease, promote wellness, and develop customized treatment options that allow patients to improve their health and wellness. It has been known that some medicines work better for some specific groups of people than others. With the advanced research on the impact of clinical, molecular, and genomic factors on diseases, it is becoming increasingly possible to develop more personalized, population cohort-based approaches for detecting, managing, and treating diseases. Predictive analytics (PA) and artificial intelligence (AI) techniques can significantly improve the speed and efficiency of analyzing vast amounts of datasets to gain actionable insights into personalized medicine. This chapter describes the four most innovative AI-powered solutions that have disrupted how personalized medicine can be implemented—digital twins (DTs), multi-omics integration, AI-driven biomarker discovery, and personalized therapeutics.

5.1 Introduction

In a move to embrace a movement in healthcare from reactive to preventive, Dr Leroy Hood introduced the concept of P4 medicine, which stands for Personalized, Predictive, Preventive, and Participatory medicine (Hood and Flores 2012). Personalized medicine refers to customizing treatment to an individual or a sub-group. Predictive medicine deals with predicting the most appropriate treatment to avoid side effects. Preventive medicine refers to individuals being well informed to recognize early signs of disease and take steps to reverse or treat it effectively. Finally, participatory medicine empowers patients to take more responsibility for their health and well-being.

Here is an old parable that can help us better understand precision/personalized medicine. The story is about seven blind men and an elephant they have never seen.

doi:10.1088/978-0-7503-2317-8ch5
5-1

Each blind man touched a different part of the elephant and described what he thought an elephant was. The first person touched the trunk and speculated that the elephant was a snake. The second person touched the tusk and said it was a plow. The third touched the ear and said it was a pal-leaf fan, and the fourth claimed the elephant to be a large water jar after touching its head. The fifth felt the torso and said it was a rock, and the sixth man vehemently said it was a tree trunk after touching the elephant's leg. The seventh man felt the tail and said it was a flywhisk. The men quarreled to ascertain their truth. The moral of the story is that we miss the big picture when considering limited knowledge as the absolute truth. However, if the seven men had collaborated and studied the elephant, they would have determined a more definitive idea of what an elephant was. Similarly, precision medicine is about combining the knowledge of the various facts that make up a human body (both internal and external) and determining what the body needs best holistically.

Now on to a more formal definition—precision/personalized medicine is defined as customizing treatment (any therapeutic measure, including diet, exercise, medications, or medical procedures) to an individual or a sub-group. The President's Council of Advisors on Science and Technology (PCAST) states that personalized medicine '... *does not literally mean the creation of drugs or medical devices that are unique to a patient, but rather the ability to classify individuals into sub-populations that differ in their susceptibility to a particular disease or their response to a specific treatment. Preventive and therapeutic interventions can then be concentrated on those who will benefit, sparing expense and side effects for those who will not*' (PCAST 2008).

Extensive analysis of a person's genetic and biological markers, social determinants of health, behaviors, or any other information provides clues on how to personalize an effective treatment and determine critical parameters, such as the duration, timing, side effects, etc. Historically, precision medicine efforts were limited by the lack of comprehensive data about individuals and the computational power required to analyze these comprehensive datasets. However, the recent advances in integrated health data collection through extensive clinical trials and real-world data (RWD) collection via wearables and remote monitoring techniques and the development of advanced big data technologies make it possible to understand much about the individual or subgroups of populations.

The chapter describes the four most AI-powered disruptive areas in the field of precision medicine—DTs, multi-omics integration, AI-driven biomarker discovery, and personalized therapeutics.

5.2 Digital twins

5.2.1 Overview

A DT, in very simple terms, is a virtual replica of an individual that integrates biological, clinical, and lifestyle data for two reasons: (1) to provide a comprehensive and real-time view of a patient's health and wellness, and (2) to simulate disease risk, diagnosis, and progression and optimize treatments. These data can come from various sources: medical records, wearable devices, imaging scans, etc. Healthcare providers can use DTs to optimize patient care by providing personalized recommendations to improve outcomes, reduce healthcare spending, and enhance overall care quality. Patients can also use their digital twins to empower themselves with knowledge about their health and wellness and receive tailored recommendations to optimally improve their quality of life. Katsoulakis *et al* (2024) define a DT for healthcare as follows: '...*we define a digital twin for healthcare as a virtual representation of a person which allows dynamic simulation of potential treatment strategy, monitoring and prediction of health trajectory, and early intervention and prevention, based on multi-scale modeling of multi-modal data such as clinical, genetic, molecular, environmental, and social factors, etc*'. They also insist that DTs should be individualized, interconnected, interactive, informative, and impactful (5Is).

The main components of a DT are a physical entity and a virtual replica, as well as a real-time bi-directional connection between the two entities to enable: (1) the physical entity (the patient) to learn from the predictions and outcomes from simulations conducted in their virtual persona, and (2) the virtual replica to learn from the real-world data from the patient.

In terms of the types of DTs, there can be organ-specific DTs, such as in cardiology, where a DT can be used to simulate heart function and predict how pacemakers and stents will perform (Baillargeon *et al* 2014) or for cardiac resynchronization (Healthineers 2025) or surgical planning (Obaid *et al* 2019). Such DTs are developed using detailed imaging data and specific computational analytics. There can be DTs for precision medicine (Shen *et al* 2024), such as disease models, where a DT is used to learn drug interactions with complex disease-based biological processes. Such DTs use molecular and clinical data.

We conducted a comprehensive literature search in PubMed using the keywords 'digital twin' and 'medicine' to pull articles from 2016 to 2025. We generated a word cloud from the titles of the articles, as shown in figure 5.1. The image displays words that appeared at least twice in the titles. The more the frequency of appearance, the

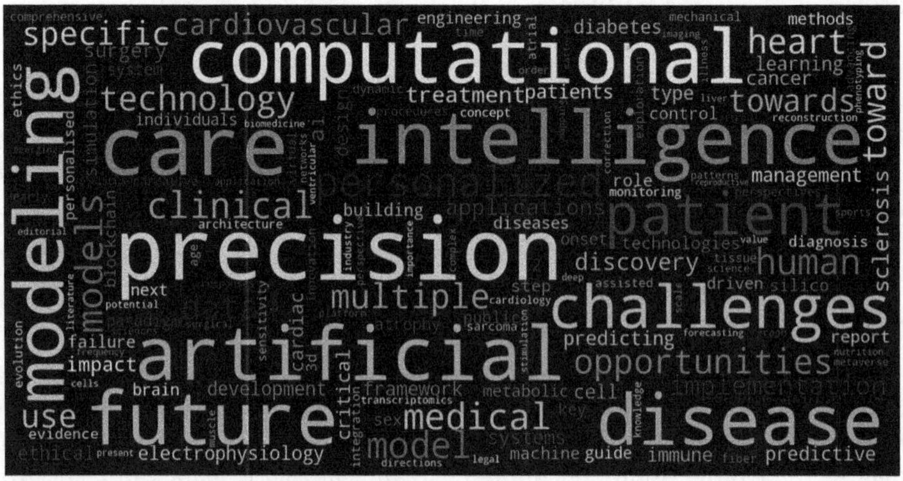

Figure 5.1. The word cloud generated from the titles of publications obtained in a comprehensive literature search in PubMed using the keywords 'digital twin' and 'medicine' to pull articles from 2016 to 2025.

bigger the font size. As is evident from the figure, *precision medicine* is a common area where DTs are helpful, and many articles focus on the computational and modeling aspects of building DTs, along with sharing challenges and opportunities.

Several review papers have explored how DTs have evolved within healthcare. One identified four main functions of DTs in healthcare management (safety, information, health and well-being, and operational control) (Elkefi and Asan 2022). Another reviewed 22 publications, summarized the various applications of DT in medicine and concluded that it is more widely used for real-time monitoring, dynamic analysis, and precise treatments (Sun *et al* 2023). Sheng *et al* (2023) conducted a quantitative review of 94 papers using a data mining method called structure topic modeling (STM). They revealed that most papers focused on technological advancements (AI, Internet of Things, etc) and application scenarios (precision, personalized healthcare, and management of public health systems). These findings align with the word cloud details (figure 5.1). Khan *et al* (2022) reviewed 18 papers to determine if any examined the use of DTs during infectious disease outbreaks such as COVID-19. Since none examined such a use case, the authors proposed a preliminary conceptual framework for using DTs for hospital management during such outbreaks. Shen *et al* (2024) conducted a systematic review to explore the benefits of DTs in improving precision health at the population level. They selected 12 studies that included patients with various conditions (cancers, Type 2 diabetes, multiple sclerosis, heart failure, qi deficiency, post-hepatectomy liver failure, and dental issues) and studied three outcome categories (personalized health management, precision individualized therapy effects, and individual risk prediction). They observed an overall effectiveness of 80% across these outcomes, highlighting the pivotal role played by DTs in advancing precision health.

Consider an example DT: Wickramasinghe *et al* (2022) proposed a DT framework for precision dementia diagnosis and care. The framework recommends that the end-user (clinician or carer) enter relevant data about the patient, such as risk factors, current symptoms, past and current treatments, progression markers, and other co-morbidities into a Decision Support System (DSS). The DSS then connects to a remote server where an AI algorithm matches the patient data with an extensive cohort of other dementia cases. If matches are found, the algorithm develops a union of best-matching cases and presents it to the end-user as a DT for the current patient. If the end-user is the clinician, he/she can use the DT to make an informed and precise diagnosis and determine personalized treatment recommendations. If it is the carer, he/she can use the presented twin information to perform assessments and obtain evidence-based care, consultation, and next-step recommendations. These details about the current patient generated by the clinician/carer will become a new entry for the DSS for future reference. This is an easily understood framework for how DTs can be used for personalized diagnosis and care for any disease.

The above example demonstrates that developing and evaluating a successful DT will involve multiple functional, technological, and operational aspects. Pellegrino *et al* (2024) provide a detailed overview of these aspects (domain, physical twin, team, sensors, data, modeling, functions, user interaction, hierarchy, status and maturity, and limitations) and their structural relationships. They then present a conceptual multi-dimensional framework for evaluating a DT in healthcare. Sel *et al* (2025a) recently examined Verification, Validation, and Uncertainty Quantification (VVUQ) for DTs in ensuring safety and efficacy, with examples in cardiology and oncology. Incorporating a VVUQ framework is important to ensure stakeholders address the challenges associated with model accuracy, computational efficiency, and the quantification of any uncertainties. The next two subsections go deeper into an organ-specific DT (heart) and a disease-specific DT (Type 2 diabetes) to help the readers understand the nuances of developing DTs.

5.2.2 Cardiac digital twins

Consider this clinical vignette. A 76-year-old woman with heart failure and other comorbid conditions such as diabetes, hypertension, and obesity is going on a walk. Her smartwatch suddenly detects an episode of atrial fibrillation and a decrease in oxygen saturation, and her wireless blood pressure cuff detects an increase in blood pressure. These abnormal biomarkers wirelessly trigger her electronic health record to run a large language model (LLM, described later) to study her medical history and clinical notes. Based on the information, it predicts that the patient is likely to experience a heart failure exacerbation. It quickly runs simulations of different medications on her outcome, compares other DTs that fit this patient's profile, and contacts her physicians with a modified treatment recommendation. The physician reviews this recommendation carefully and contacts the patient about the next plan. In the meantime, the LLM updates its training data with this current episode to prepare it for accurate future predictions. This vignette was presented in an article by Thangaraj *et al* (2024) where they describe how generative AI tools can aid cardiovascular care. They also describe the various other ways DTs can be used in cardiology—evidence generation (DTs as controls in clinical trials), generating synthetic cardiac data (for simulating multiple patient populations by building a synthetic cohort and studying treatment effects in these populations), for granular phenotyping of disorders such as cardiomyopathies, for coronary and structural heart simulations.

DTs in cardiology can be built using mechanistic, statistical, or hybrid models. Mechanistic models use electrophysiological parameters, fluid dynamics principles, etc, to characterize cardiac physiology. For example, Gillette *et al* (2021) developed a framework for generating DTs of cardiac electrophysiology using electrocardio-grams (ECGs). They created replicas of the heart using 3D magnetic resonance imaging (MRI) scans (anatomical twinning) and simulated cardiac electrophysiol-ogy (functional twinning) using clinical ECG measurements and ECG modeling. They demonstrated near real-time simulation of cardiac electrophysiology using this DT. Statistical models, however, use probabilistic methods, such as deep learning, generative AI, etc, to utilize multi-modal data to learn patterns across patient populations to predict disorders and analyze longitudinal information (Coorey *et al* 2022). Sel *et al* (2025b) review current cardiovascular system modeling techniques and also highlight future directions and research opportunities in developing and using cardiac DTs.

5.2.3 Digital twins for Type 2 diabetes

Diabetes is a complex chronic condition that can be managed successfully by the constant monitoring of blood glucose levels and medication management, in addition to behavioral changes in diet, exercise, and stress management. DTs are well suited for managing diabetes as they can incorporate and track all relevant biomarkers. Sarani Rad *et al* (2024) recently proposed a holistic, patient-centric diabetes management framework that uses personal health knowledge graphs (PHKGs) to capture the complex relationships between the diverse biomarkers related to the disease. They have also explained how their approach can help diabetes patients in four ways:

1. Personalized blood glucose regulation: Here, they use the DT to train and test insulin optimization strategies. The approach used reinforcement learning to enable personalized insulin optimization, enhance blood glucose control, and reduce hyper and hypoglycemic risks (Sarani Rad and Li 2023).
2. Glucose prediction: This DT was able to generate individual glucose forecasts to empower users to take steps to maintain their glucose levels within the optimal range (Yang and Li 2023).
3. Healthcare data DT explorer: This DT uses the relationship between PHKGs and health concepts to empower users in data literacy so they can comprehend their health information easily at their own pace (Hendawi and Li 2024).
4. Personalized meal recommendation: The PHKG collects data on patient dietary preferences and allergies. These data points, along with other clinical and treatment information, are used as input in this meal recommendation engine (Amiri *et al* 2023).

Zhang *et al* (2024a) also introduced a DT framework for Type 2 diabetes that utilized machine learning algorithms, multi-omic and clinical datasets, knowledge graphs, and mechanistic models. Studies have also analyzed whether commercially available DT interventions can significantly improve a patient's glycemic control, reduce the intake of anti-diabetic medications, and improve metabolic health (Shamanna *et al* 2024). The users in their program were given a continuous glucose monitor (CGM), activity tracker (Fitbit Charge), smart scale (Powermax BCA-130), and access to the smartphone app—Whole Body Digital Twin™ (Twin Health 2025). Clinical history and lab parameters were also obtained. Users entered daily food logs. All these variables collected accounted for about 200 features that were then used to train AI models to predict post-prandial glucose response. The study also provided interventions via AI-driven nudges to encourage better diet and lifestyle habits. Their results indicate that DTs can have a beneficial role in diabetes management. Thamotharan *et al* (2023) proposed a human DT framework based on several mathematical and deep learning-based models for managing Type 2 diabetes in elderly patients. It is a detailed paper that interested readers should check out to design a DT for diabetes management.

Readers interested in learning more about DT application areas can read the article by Meijer *et al* (2023), which reviews the methodological development of

DDT solutions in several other areas of healthcare—artificial pancreas, single-cell flux analysis, protein and DNA interactions, clinical reporting in oncology, predicting drug effectiveness, drug repurposing, etc

5.2.4 Challenges

It is important to note that the success of DTs relies on two important assumptions:
1. *The first is that computational models can be developed to predict a patient's response to a specific treatment.* Humans have individual histories of host-microbiome and host–pathogen interactions, which can easily impact the accuracy of predictive models. It is also nearly impossible to train models on full individual exposomes, which includes all past exposure to various environmental factors such as diet and social situations (De Domenico *et al* 2025).
2. *The second is that heterogeneous data sources are available to build massive databases from various populations, enough to split them into subgroups that have distinctive clinical features.* Even though we have made advancements in collecting data from wearables, there are still challenges associated with the consistent use of wearables by patients. There are still no standardized practices to collect multi-omics, multi-modal datasets from individuals easily. For example, many health conditions are dependent on the immune system. So, it is important to include immune system-related data when designing DTs. However, the immune response is complex, and its modeling requires deep collaboration among clinical, immunological, and computation modeling groups. Niarakis *et al* (2024) go into the details of the progress made in developing immune DTs and outline important aspects of designing an immune DT and prerequisites for implementation.

Several recent studies acknowledge the methodological challenges and opportunities of developing DTs in medicine (Meijer *et al* 2023, Zhang *et al* 2024b, De Domenico *et al* 2025). A more comprehensive approach is needed to design effective DTs, considering the multi-scale nature of human beings. Such an approach should first determine which key biological, clinical, and environmental variables are needed to simulate a physical entity adequately. Then comes the issue of data collection. As described earlier, the primary challenge is obtaining the different types of omics data required to develop a fully comprehensive medical DT. For DT models that rely on monitoring health in real-time, continuous development of noninvasive, high-throughput data collection methods should be in place to ensure the accuracy and effectiveness of the model. Once the data is in place, next comes modeling. Machine learning-based predictive models can struggle because they are difficult to generalize to situations they have not been trained on. However, recent advances in generative AI, embodied AI, agentic AI, and the metaverse hold immense promise in handling multi-modal, multi-scale large datasets with ease (Zhang *et al* 2024a, 2024b).

LLMs, a type of generative AI, are deep neural networks trained on large amounts of data with billions of parameters to generate human-like text responses. Multi-modal LLMs, as the name suggests, take in diverse data. Embodied AI are models that learn from interactions with environments. They have been particularly useful in mental health (Fiske *et al* 2019) and medical robotics (De Micco *et al* 2024). Such systems use machine learning, computer vision, and other technologies to create AI systems that can perceive, act, and collaborate. LLMs plus embodied AI systems, often referred to as AI agents, have also been explored (Zhou *et al* 2023). While generative AI focuses on creating content-like text, images, or code based on existing patterns, agentic AI aims to create autonomous agents that can perform tasks and make decisions autonomously. This agentic AI might allow for more dynamic and realistic modeling of complex human systems, resulting in improved optimization and prediction capabilities compared to traditionally built digital twins that rely on static data analysis. Some studies have taken the next step to combine DTs with AI agents (Croatti *et al* 2020) for managing severe trauma cases. They mirrored the trauma management process by two DTs, one modeling the pre-hospital phase, where emergency services take care of the patient at the scene of the incident, and one modeling the trauma phase, where the patient arrives at the hospital and is now in the care of the trauma team. These DTs are overseen by agentic AI, which continuously learns from the trauma team's actions and acts as a physician assistant, supporting the physician with timely documentation, etc. Sounds exciting?

Add to all these advancements—the metaverse. It represents a collective virtual, immersive, digital environment created by technologies such as augmented reality (AR) and virtual reality (VR). Healthcare providers and patients can interact in this simulated space in real time. For example, multidisciplinary teams of providers can collaborate in this space to remotely monitor a patient's DT and make more informed and effective treatment decisions (Yang *et al* 2022). Patients can also benefit from these empowering environments that inform them about their care using patient engagement and education tools (Chengoden *et al* 2023).

On the challenging side, these advanced technologies require tremendous computing power. While acquiring data from large populations, issues and concerns related to data privacy, security, ethics, and confidentiality must also be addressed. Considering the opportunities and challenges of developing medical DTs, do the benefits outweigh the risks?

We will have to wait and watch.

5.3 Multi-omics integration

5.3.1 Overview

Recently, the National Institutes of Health awarded $50.3 million for multi-omics research on health and disease. The use of multi-omics datasets strengthens DT development. Multi-omics data combines data from genomics, epigenomics, transcriptomics, proteomics, metabolomics, and other omics disciplines. Here is a quick overview of what these different -omics stand for.

- *Genomics:* It investigates the structure, function, and mapping of information codes in our DNA (genes). Such investigations have led to understanding how these genes mutate in the event of a disease.

- *Epigenomics:* It is the study of epigenetic modifications, such as DNA methylation and histone modifications, that affect the gene expression in a cell. It is an interesting science that explains why identical twins with the same DNA sequence have different characteristics. The epigenome can change based on the environment and, therefore, could be valuable data to study chronic diseases such as cancer, metabolic syndromes, and cardiovascular disease.
- *Transcriptomics:* It is the study of the transcriptome, a collection of RNA in a cell, tissue, or organ at a given time. This study can indicate which genes are active at any given time and help understand the structure of genes. It bridges genes and proteins and can inform much about human biology.
- *Proteomics:* As the name indicates, proteomics is the study of proteins' structure, function, and interactions. It gives us important insights into how proteins maintain metabolic processes and regulate gene expression. It is highly useful in the early detection of diseases, helps identify target molecules for drug discovery, and tailors personalized medical treatments.
- *Metabolomics:* The study of small molecules called metabolites produced by the body during the breakdown of food, drugs, and chemicals. It can identify biomarkers for diseases such as cancer and cardiometabolic diseases.

Integrating large volumes of multi-omics data accurately to derive value from them is challenging. With the increase in the scale of such biological data and the advancements in processing technologies such as generative and agentic AI tools, it is now possible to more accurately understand human biology and improve patient outcomes. Multi-omics data integration methods can be classified into early, intermediate, and late (Cai *et al* 2022). In **early integration**, the features from all the different datasets are concatenated. However, if the patient sample size is small, features exceed patients and might lead to the 'curse of dimensionality' problem. Data heterogeneity issues must also be addressed as different omics datasets have different distributions and types. In the **intermediate integration** strategy, each dataset is transformed independently into a simpler representation. For example, data dimensionality could be reduced, and data heterogeneity issues could also be addressed. Advanced AI algorithms can then combine and analyze the transformed representations of several omics datasets. In **late integration**, each omics dataset is analyzed by omic-specific AI algorithms, and the results/predictions from each omics dataset are combined at the end. The disadvantage of this approach is that inter-omics integration-related insights are not captured in the results.

Various algorithms for these integrations exist. Some of them are multiset sparse partial least squares path modeling (msPLS) (Csala *et al* 2020) and MOTA: network-based multi-omic data integration for biomarker discovery (Fan *et al* 2020). Concerning software options, some of them are: mixOmics (an R package with various multivariate methods for integrating different omics datasets), OmicsNet (a web-based platform for multi-omics integration and network visualization), CDIAM Multi-Omics Studio (a comprehensive platform for analyzing and integrating multiple omics data types), Holomics (a user-friendly R Shiny application for multi-omics data integration), and MiBiOmics (an interactive web application for multi-omics data visualization and integration).

5.3.2 Clinical applications

Below are some clinical applications of multi-omics integration.

- *Pregnancy:* Analyzing longitudinal data collected during the various stages of pregnancy can help predict risks of complications such as preterm birth, preeclampsia, and fetal growth restriction early enough to address them quickly (Pammi *et al* 2023). Ghaemi *et al* (2019) analyzed immunome, transcriptome, microbiome, proteome, and metabolome data from 17 pregnant women. They observed a chronology of biologically diverse events during pregnancy and novel interactions between different biological modalities. Marić *et al* (2022) analyzed six omics datasets from a cohort of pregnant women to develop a model for early prediction of *preeclampsia*. In another study, more than 7000 plasma analytes and peripheral immune cell responses were analyzed to predict the time to *spontaneous labor*. They observed key alterations in the maternal metabolome, proteome, and immunome that led to a molecular shift from pregnancy to pre-labor biology 2–4 weeks before delivery (Stelzer *et al* 2021). A recent study by Jehan *et al* (2020) describes how cell-free transcriptomics, urine metabolomics, and plasma proteomics can be used to identify the biological measurements associated with preterm birth.
- *Neonatal intensive care unit (NICU)-related diseases:* AI-powered multi-omics data analysis can potentially shed light on poorly defined NICU diseases such as sepsis, necrotizing enterocolitis, bronchopulmonary dysplasia, and various forms of encephalopathy (Pammi *et al* 2023).
- *Kidney disease:* Kidney disease is known for its intricate pathogenic mechanisms and a lack of precise therapeutic modalities based on molecular pathologies. Multi-omics data-based analysis could aid in understanding the underlying biological patterns of this disease, which is currently one of the major diseases affecting many globally. Liu *et al* (2024) and Delrue and Speeckaert (2024) have attempted to find strategies to integrate multi-omics data and machine learning algorithms to understand this disease better.
- *Oncology:* Radiomics, referring to quantitative features extracted from clinical imaging, has been extensively studied in cancer research. Wei *et al* (2023) present how radiomics and other molecular biomarkers are used in precision oncology. Their review also discusses AI-based modeling methods and presents examples of how multi-omics research can help improve cancer subtyping, risk stratification, prognostication, prediction, and clinical decision-making.
- *Cardiovascular disease (CVD):* Sopic *et al* (2023) present a detailed analysis of how AI tools can analyze multi-omics data and clinical datasets to develop preventive and therapeutic strategies for atherosclerotic CVD management. Wang *et al* (2023) present how multi-omics technologies, including bulk

omics and single-cell omics technologies, can contribute to precision medicine. They specifically highlight network medicine-based integration of multi-omics data for developing therapeutics for CVD management.

5.4 AI-driven biomarker discovery

5.4.1 Overview

Biomarker identification is key for advancing medical science and improving disease detection, management, and treatment outcomes. Some of the key reasons why we need to identify these biomarkers are listed below:

- to identify diseases early (e.g., prostate-specific antigen for prostate cancer screening),
- to tailor treatments to individual patients or subgroups based on their unique biology (e.g., breast cancer patients with the HER2 biomarker can benefit from HER2-targeted therapies),
- to classify diseases and their subtypes to enable targeted interventions (e.g., molecular subtypes of lung cancer),
- to predict risks and prognosis (e.g., the presence of BRCA1 and BRCA2 mutations predict a higher risk of developing breast and ovarian cancers),
- to monitor disease progression in real-time (e.g., HbA1c levels monitoring in diabetes patients),
- to develop new drugs and therapies, to select appropriate patient populations for clinical trials (e.g., using PD-L1 expression for immunotherapy trials), to monitor drug safety and efficacy in trials,
- to understand the underlying molecular and cellular mechanisms of diseases (e.g., understanding Alzheimer's disease via biomarkers like beta-amyloid and tau proteins,
- to reduce healthcare costs by enabling targeted interventions and avoiding unnecessary treatments, and
- to enable personalized health monitoring (dynamic biomarkers such as heart rate variability, glucose data, etc, from wearables).

Multi-omics data integration will help us better understand the functional interactions in human biology. It will also help discover novel meta-biomarkers that can help profile disease prognosis and prediction. AI can more easily help detect these biomarkers by analyzing volumes of omics datasets and radiomics data, pathology, and microbiome datasets, as described in the previous sections. One example is the development of PandaOmics software, an AI-driven platform for discovering therapeutic targets and biomarkers (Kamya *et al* 2024).

5.4.2 Clinical applications

Below are some clinical applications of AI-driven biomarker discovery.

- *Oncology:* Prelaj *et al* (2024) conducted a systematic review of 90 studies using genomics, transcriptomics, epigenomics, radiomics, digital pathology, and real-world and multimodality datasets to study the efficacy of immune checkpoint inhibitors in cancer patients. They confirmed that most studies utilized AI to discover novel predictive biomarkers. However, at this stage, the findings are just hypothesis-generating insights that are not yet suitable for direct clinical implementation. Large-scale trials are needed to establish the efficacy of these biomarkers. The other issue with the discovery of such biomarkers and the resulting targeted therapies is the cost involved in the development process. High cost limits the accessibility to patients all around the world, exacerbating socioeconomic disparities. To address this less-addressed issue, Ligero *et al* (2025) reviewed studies that used the latest AI technologies, such as LLMs, to support cancer treatment decisions with cost-effective biomarkers. They also highlight how AI-based biomarker discovery can aid in detecting biomarkers in a less-researched area—prediction treatment response and prognosis. For this use case, biomarkers can come from the following modalities.

- ○ Genome analysis. For example, Elmarakeby *et al* (2021) developed P-NET, a deep learning model using molecular data from genomic information to stratify prostate cancer patients based on their treatment resistance.
- ○ Medical imaging. For example, several studies have developed AI-based algorithms to stratify patients using digitized hematoxylin and eosin slides to optimize treatment and therapy decisions (Zeng *et al* 2023, Ahn *et al* 2024).
- ○ Electronic health records. Patient clinical history, including pre-existing diseases and demographics, can be critical in determining treatment response (Huang *et al* 2020).

- *Cardiovascular disease:* Because of the diversity of CVDs and their varying intricate genetic composition, the detection and development of personalized interventions could benefit cardiology. DeGroat *et al* (2024) proposed a novel approach using statistics and AI to identify cardiac biomarkers by analyzing complete transcriptomes of CVD patients. They uncovered 18 biomarkers that could predict CVD with up to 96% accuracy. A review by Ghantous *et al* (2020) summarized novel, promising, early-stage protein and miRNA biomarker options and existing biomarkers for detecting hypertension and other cardiovascular diseases. They recommend that a multi-marker panel strategy might be more promising and effective in risk stratification and classification of patients with CVD.

- *Alzheimer's disease and dementia:* As with many diseases, early diagnosis and accurate subtyping of disease could improve the effectiveness of dementia management. Winchester *et al* (2023) have summarized how AI has powered the discovery of biomarkers for dementia.

Despite the rapid advancements in using AI for biomarker discovery, several limitations exist. The three key ones are listed below.

1. Comprehensive evaluation using large-scale trials is still lacking. Even though many AI-based tools have received FDA approval, prospective clinical trial evaluations are still lacking (Prelaj *et al* 2024).

2. AI-based biomarker detection methodologies are not always economically viable in low-income countries or smaller healthcare centers. Economic feasibility studies are also lacking.
3. A well-known limitation of AI-based models is the possibility of bias in algorithmic development.

Ligero *et al* (2025) present some strategies for addressing these key challenges in using AI for biomarker discovery.

5.5 Personalized therapeutics

Most research in AI in precision oncology has primarily focused on drug development. Of late, many groups are studying how AI can be used to develop personalized drug and cell therapies (Taherdoost and Ghofrani 2024). Digital twin development, multi-omics data integration, biomarker discovery, and personalized therapeutics development are all exciting areas advancing the field of precision medicine. The common thread advancing these areas is AI. The previous sections extensively describe the opportunities and challenges of using AI to advance these areas. With respect to the development of personalized therapeutics, AI can play several key roles:

- *Drug discovery:* Drug discovery and development are tedious and time-consuming, and AI can help speed the process. It can help analyze molecular structures, verify drug candidates, and understand disease mechanisms. Zhang *et al* (2025) reviewed the recent advancements in AI applications across the entire drug development workflow—disease target identification, drug discovery, preclinical and clinical studies, and post-market surveillance. AI platforms like BenevolentAI use deep learning to predict drug–protein interactions and suggest novel therapeutic targets (Benevolent AI 2025).
- *Identifying personalized therapeutic targets:* As highlighted earlier, AI techniques can analyze large multi-omics datasets to identify unique molecular signatures/biomarkers in patients that specific drugs can target (You *et al* 2022, Shams 2024).
- *Predicting drug response:* A key factor in the effectiveness of a drug is predicting how an individual will respond to that drug based on their unique genetic, molecular, clinical, and lifestyle factors (Lin *et al* 2024). AI can be used to build pharmacogenomics models that use genetic data to predict drug response.
- *Precision gene and cell therapies:* AI can identify the best candidates for gene-editing therapies like CRISPR-Cas9 based on patient-specific genetic profiles (Dixit *et al* 2024).
- *Identifying therapeutic pathways for rare diseases:* Diseases such as cystic fibrosis and Huntington's have been better understood because AI has been used to study patients' genetic data and identify novel therapeutic pathways (Amaral and Harrison 2023, Ganesh *et al* 2023, Cheng *et al* 2024).
- *Predicting adverse drug reactions:* Detecting unintended adverse drug reactions in pharmacological research is important. AI algorithms, such as the one proposed by Li *et al* (2024), have proven highly accurate in this area.

A key bottleneck in translating precision care therapies from bench to bedside is the lack of speed in getting regulatory clearance). Derraz *et al* (2024) have explored emerging concepts and new ideas for regulating AI-enabled personalized cancer therapies in the context of existing and in-development governance frameworks.

5.6 Conclusions

This chapter provided an overview of the latest advancements in precision medicine and how AI has fast-tracked this evolution. It also described the challenges that must be addressed before mainstream implementation and use. As research and development improve disease stratification according to an individual's genetic makeup and optimize treatments, solutions should be developed to empower individuals to establish a wellness baseline, monitor the progression from wellness to disease state, and monitor treatments. The key focus in healthcare will be wellness. More efficient workflow protocols, data pipelines, and comprehensive mobile and web apps should be developed to allow providers to longitudinally follow each patient and detect any disease states long before the onset of disease symptoms. As we march forward in this revolutionary era of medicine, more efforts will be required to educate patients, physicians, and the healthcare community to embrace these changes. EdTech solutions should be developed to educate this mostly conservative community and help improve the adoption of health tech solutions.

Dr Leroy Hood hopes and anticipates that in a few years, we may be able to develop and use a small hand-held device that can prick a finger and make 2500 blood measurements to detect organ-specific proteins (Hood and Flores 2012). The data can be used to detect diseases much earlier. However, to speed up research and to continue making advancements, all this data should be made available to investigators worldwide after appropriate anonymization and strong protections against unethical exploitation.

References

Ahn B, Moon D, Kim H S *et al* 2024 Histopathologic image-based deep learning classifier for predicting platinum-based treatment responses in high-grade serous ovarian cancer *Nat. Commun.* **15** 4253

Amaral M D and Harrison P T 2023 Development of novel therapeutics for all individuals with CF (the future goes on) *J. Cyst. Fibros.* **22** S45–9

Amiri M, Li J and Hasan W 2023 Personalized flexible meal planning for individuals with diet-related health concerns: system design and feasibility validation study *JMIR Form. Res.* **7** e46434

Baillargeon B, Rebelo N, Fox D D, Taylor R L and Kuhl E 2014 The living heart project: a robust and integrative simulator for human heart function *Eur. J. Mech. A. Solids* **48** 38–47

Benevolent AI 2025 https://benevolent.com/

Cai Z, Poulos R C, Liu J and Zhong Q 2022 Machine learning for multi-omics data integration in cancer *iScience* **25** 103798

Cheng Y, Zhang S and Shang H 2024 Latest advances on new promising molecular-based therapeutic approaches for Huntington's disease *J. Transl. Int. Med.* **12** 134–47

Chengoden R *et al* 2023 Metaverse for healthcare: a survey on potential applications, challenges and future directions *IEEE Access* **11** 12765–95

Coorey G, Figtree G A, Fletcher D F *et al* 2022 The health digital twin to tackle cardiovascular disease—a review of an emerging interdisciplinary field *NPJ Digit. Med.* **5** 126

Croatti A, Gabellini M, Montagna S *et al* 2020 On the integration of agents and digital twins in healthcare *J. Med. Syst.* **44** 161

Csala A, Zwinderman A H and Hof M H 2020 Multiset sparse partial least squares path modeling for high dimensional omics data analysis *BMC Bioinf.* **21** 9

De Domenico M, Allegri L, Caldarelli G *et al* 2025 Challenges and opportunities for digital twins in precision medicine from a complex systems perspective *NPJ Digit. Med.* **8** 37

DeGroat W, Abdelhalim H, Peker E *et al* 2024 Multimodal AI/ML for discovering novel biomarkers and predicting disease using multi-omics profiles of patients with cardiovascular diseases *Sci. Rep.* **14** 26503

De Micco F, Tambone V, Frati P, Cingolani M and Scendoni R 2024 Disability 4.0: bioethical considerations on the use of embodied artificial intelligence *Front. Med. (Lausanne)* **11** 1437280

Delrue C and Speeckaert M M 2024 Decoding kidney pathophysiology: omics-driven approaches in precision medicine *J. Pers. Med.* **14** 1157

Derraz B, Breda G, Kaempf C *et al* 2024 New regulatory thinking is needed for AI-based personalised drug and cell therapies in precision oncology *NPJ Precis. Oncol.* **8** 23

Dixit S, Kumar A, Srinivasan K, Vincent P M D R and Ramu Krishnan N 2024 Advancing genome editing with artificial intelligence: opportunities, challenges, and future directions *Front. Bioeng. Biotechnol.* **11** 1335901

Elkefi S and Asan O 2022 Digital twins for managing health care systems: rapid literature review *J. Med. Internet Res.* **24** e37641

Elmarakeby H A, Hwang J, Arafeh R *et al* 2021 Biologically informed deep neural network for prostate cancer discovery *Nature* **598** 348–52

Fan Z, Zhou Y and Ressom H W 2020 MOTA: network-based multi-omic data integration for biomarker discovery *Metabolites* **10** 144

Fiske A, Henningsen P and Buyx A 2019 Your robot therapist will see you now: ethical implications of embodied artificial intelligence in psychiatry, psychology, and psychotherapy *J. Med. Internet Res.* **21** e13216

Ganesh S, Chithambaram T, Krishnan N R, Vincent D R, Kaliappan J and Srinivasan K 2023 Exploring Huntington's disease diagnosis via artificial intelligence models: a comprehensive review *Diagnostics (Basel)* **13** 3592

Ghaemi M S, DiGiulio D B, Contrepois K *et al* 2019 Multiomics modeling of the immunome, transcriptome, microbiome, proteome and metabolome adaptations during human pregnancy *Bioinformatics* **35** 95–103

Ghantous C M, Kamareddine L, Farhat R *et al* 2020 Advances in cardiovascular biomarker discovery *Biomedicines* **8** 552

Gillette K, Gsell M A F, Prassl A J *et al* 2021 A framework for the generation of digital twins of cardiac electrophysiology from clinical 12-leads ECGs *Med. Image Anal.* **71** 102080

Healthineers S 2025 The Value of Digital Twin Technology https://siemens-healthineers.com/en-us/services/value-partnerships/asset-center/white-papers-articles/value-of-digital-twin-technology

Hendawi R and Li J 2024 Comprehensive personal health knowledge graph for effective management and utilization of personal health data *2024 IEEE First Int. Conf. on*

Artificial Intelligence for Medicine, Health and Care (AIMHC) (Laguna Hills, CA, USA) pp 92–100

Hood L and Flores M 2012 A personal view on systems medicine and the emergence of proactive P4 medicine: predictive, preventive, personalized and participatory *Nat. Biotechnol.* **29** 613–24

Huang S C, Pareek A, Seyyedi S, Banerjee I and Lungren M P 2020 Fusion of medical imaging and electronic health records using deep learning: a systematic review and implementation guidelines *NPJ Digit. Med.* **3** 136

Jehan F *et al* 2020 Multiomics characterization of preterm birth in low- and middle-income countries [published correction appears in *JAMA Netw. Open* 2021 Feb 1;4(2):e210399. 10.1001/jamanetworkopen.2021.0399] *JAMA Netw. Open* **3** e2029655

Kamya P, Ozerov I V, Pun F W *et al* 2024 PandaOmics: an AI-driven platform for therapeutic target and biomarker discovery *J. Chem. Inf. Model.* **64** 3961–9

Katsoulakis E, Wang Q, Wu H *et al* 2024 Digital twins for health: a scoping review *NPJ Digit. Med.* **7** 77

Khan A, Milne-Ives M, Meinert E, Iyawa G E, Jones R B and Josephraj A N 2022 A scoping review of digital twins in the context of the COVID-19 pandemic *Biomed. Eng. Comput. Biol.* **13** 11795972221102115

Li S, Zhang L, Wang L *et al* 2024 BiMPADR: a deep learning framework for predicting adverse drug reactions in new drugs *Molecules* **29** 1784

Ligero M, El Nahhas O S M, Aldea M and Kather J N 2025 Artificial intelligence-based biomarkers for treatment decisions in oncology *Trends Cancer.* **11** P232–44

Lin C X, Guan Y and Li H D 2024 Artificial intelligence approaches for molecular representation in drug response prediction *Curr. Opin. Struct. Biol.* **84** 102747

Liu X, Shi J, Jiao Y *et al* 2024 Integrated multi-omics with machine learning to uncover the intricacies of kidney disease *Brief. Bioinform.* **25** bbae364

Marić I, Contrepois K, Moufarrej M N *et al* 2022 Early prediction and longitudinal modeling of preeclampsia from multiomics *Patterns (NY)* **3** 100655

Meijer C, Uh H W and El Bouhaddani S 2023 Digital twins in healthcare: methodological challenges and opportunities *J. Pers. Med.* **13** 1522

Niarakis A, Laubenbacher R, An G *et al* 2024 Immune digital twins for complex human pathologies: applications, limitations, and challenges *NPJ Syst. Biol. Appl.* **10** 141

Obaid D R, Smith D, Gilbert M, Ashraf S and Chase A 2019 Computer simulated 'Virtual TAVR' to guide TAVR in the presence of a previous Starr-Edwards mitral prosthesis *J Cardiovasc Comput. Tomogr.* **13** 38–40

Pammi M, Aghaeepour N and Neu J 2023 Multiomics, artificial intelligence, and precision medicine in perinatology *Pediatr Res.* **93** 308–15

PCAST 2008 https://hsdl.org/c/presidents-council-releases-findings-on-new-class-of-medicine/

Pellegrino G, Gervasi M, Angelelli M *et al* 2024 A conceptual framework for digital twin in healthcare: evidence from a systematic meta-review *Inform. Syst. Front.* **27** 7–32

Prelaj A, Miskovic V, Zanitti M *et al* 2024 Artificial intelligence for predictive biomarker discovery in immuno-oncology: a systematic review *Ann. Oncol.* **35** 29–65

Sarani Rad F and Li J 2023 Optimizing blood glucose control through reward shaping in reinforcement learning *Proc. of the IEEE Int. Conf. on E-Health Networking, Application and Services (HealthCom) (Chongqing, China)* pp 342–7

Sarani Rad F, Hendawi R, Yang X and Li J 2024 Personalized diabetes management with digital twins: a patient-centric knowledge graph approach *J. Pers. Med.* **14** 359

Sel K, Hawkins-Daarud A, Chaudhuri A *et al* 2025a Survey and perspective on verification, validation, and uncertainty quantification of digital twins for precision medicine *NPJ Digit. Med.* **8** 40

Sel K, Osman D, Zare F *et al* 2025b Building digital twins for cardiovascular health: from principles to clinical impact *J. Am. Heart Assoc.* **13** e31981

Shamanna P, Erukulapati R S, Shukla A *et al* 2024 One-year outcomes of a digital twin intervention for type 2 diabetes: a retrospective real-world study *Sci. Rep.* **14** 25478

Shams A 2024 Leveraging state-of-the-art AI algorithms in personalized oncology: from transcriptomics to treatment *Diagnostics (Basel)* **14** 2174

Shen M D, Chen S B and Ding X D 2024 The effectiveness of digital twins in promoting precision health across the entire population: a systematic review *NPJ Digit. Med.* **7** 145

Sheng B, Wang Z, Qiao Y, Xie S Q, Tao J and Duan C 2023 Detecting latent topics and trends of digital twins in healthcare: a structural topic model-based systematic review *Digit. Health* **9** 20552076231203672

Sopic M, Vilne B, Gerdts E *et al* 2023 Multiomics tools for improved atherosclerotic cardiovascular disease management *Trends Mol. Med.* **29** 983–95

Stelzer I A, Ghaemi M S, Han X *et al* 2021 Integrated trajectories of the maternal metabolome, proteome, and immunome predict labor onset *Sci. Transl. Med.* **13** eabd9898

Sun T, He X and Li Z 2023 Digital twin in healthcare: recent updates and challenges *Digit. Health* **9** 20552076221149651

Taherdoost H and Ghofrani A 2024 AI's role in revolutionizing personalized medicine by reshaping pharmacogenomics and drug therapy *Intell. Pharm.* **2** 643–50

Thamotharan P, Srinivasan S, Kesavadev J *et al* 2023 Human digital twin for personalized elderly type 2 diabetes management *J. Clin. Med.* **12** 2094

Thangaraj P M, Benson S H, Oikonomou E K, Asselbergs F W and Khera R 2024 Cardiovascular care with digital twin technology in the era of generative artificial intelligence *Eur. Heart J.* **45** 4808–21

Twin Health 2025 https://usa.twinhealth.com/

Wang R S, Maron B A and Loscalzo J 2023 Multiomics network medicine approaches to precision medicine and therapeutics in cardiovascular diseases *Arterioscler. Thromb. Vasc. Biol.* **43** 493–503

Wei L, Niraula D, Gates E D H *et al* 2023 Artificial intelligence (AI) and machine learning (ML) in precision oncology: a review on enhancing discoverability through multiomics integration *Br. J. Radiol.* **96** 20230211

Wickramasinghe N, Ulapane N, Andargoli A *et al* 2022 Digital twins to enable better precision and personalized dementia care *JAMIA Open* **5** ooac072

Winchester L M, Harshfield E L, Shi L *et al* 2023 Artificial intelligence for biomarker discovery in Alzheimer's disease and dementia *Alzheimers Dement.* **19** 5860–71

Yang D, Zhou J, Chen R *et al* 2022 Expert consensus on the metaverse in medicine *Clin. eHealth* **5** 1–9

Yang X and Li J 2023 Edge AI empowered personalized privacy-preserving glucose prediction with federated deep learning *2023 IEEE Int. Conf. on E-health Networking, Application and Services (Healthcom) (Chongqing, China)* pp 224–30

You Y, Lai X, Pan Y *et al* 2022 Artificial intelligence in cancer target identification and drug discovery *Signal Transduct. Target. Ther.* **7** 156

Zeng Q, Klein C, Caruso S *et al* 2023 Artificial intelligence-based pathology as a biomarker of sensitivity to atezolizumab-bevacizumab in patients with hepatocellular carcinoma: a multi-centre retrospective study *Lancet Oncol.* **24** 1411–22

Zhang Y, Qin G, Aguilar B *et al* 2024a A framework towards digital twins for type 2 diabetes *Front. Digit. Health* **6** 1336050

Zhang K, Zhou H Y, Baptista-Hon D T *et al* 2024b Concepts and applications of digital twins in healthcare and medicine *Patterns (NY)* **5** 101028

Zhang K, Yang X, Wang Y *et al* 2025 Artificial intelligence in drug development *Nat. Med.* **31** 45–59

Zhou H Y, Yu Y, Wang C *et al* 2023 A transformer-based representation-learning model with unified processing of multimodal input for clinical diagnostics *Nat. Biomed. Eng.* **7** 743–55

www.ingramcontent.com/pod-product-compliance
Lightning Source LLC
Chambersburg PA
CBHW082105210326
41599CB00033B/6586